The Future by the Sea: Engaging with Maritime Archaeological Research during the Climate Emergency

AF631530

About Access Archaeology

Access Archaeology offers a different publishing model for specialist academic material that might traditionally prove commercially unviable, perhaps due to its sheer extent or volume of colour content, or simply due to its relatively niche field of interest. This could apply, for example, to a PhD dissertation or a catalogue of archaeological data.

All *Access Archaeology* publications are available as a free-to-download pdf eBook and in print format. The free pdf download model supports dissemination in areas of the world where budgets are more severely limited, and also allows individual academics from all over the world the opportunity to access the material privately, rather than relying solely on their university or public library. Print copies, nevertheless, remain available to individuals and institutions who need or prefer them.

The material is refereed and/or peer reviewed. Copy-editing takes place prior to submission of the work for publication and is the responsibility of the author. Academics who are able to supply print-ready material are not charged any fee to publish (including making the material available as a free-to-download pdf). In some instances the material is type-set in-house and in these cases a small charge is passed on for layout work.

Our principal effort goes into promoting the material, both the free-to-download pdf and print edition, where *Access Archaeology* books get the same level of attention as all of our publications which are marketed through e-alerts, print catalogues, displays at academic conferences, and are supported by professional distribution worldwide.

The free pdf download allows for greater dissemination of academic work than traditional print models could ever hope to support. It is common for a free-to-download pdf to be downloaded hundreds or sometimes thousands of times when it first appears on our website. Print sales of such specialist material would take years to match this figure, if indeed they ever would.

This model may well evolve over time, but its ambition will always remain to publish archaeological material that would prove commercially unviable in traditional publishing models, without passing the expense on to the academic (author or reader).

Archaeopress Access Archaeology

The Future by the Sea

Engaging with Maritime Archaeological Research during the Climate Emergency

Edited by
Katerina Velentza and Brandon Braun

Access Archaeology

Archaeopress Access Archaeology

Archaeopress Publishing Ltd
13-14 Market Square
Bicester
Oxfordshire OX26 6AD
United Kingdom

www.archaeopress.com

ISBN 978-1-80583-294-2
ISBN 978-1-80583-295-9 (e-Pdf)

© the individual authors and Archaeopress 2026

Cover: View of the the Gorrondatxe boat partially excavated. © J.M. Matés Luque.

All rights reserved. No part of this book may be reproduced, or transmitted, in any form or by any means, electronic, mechanical, photocopying or otherwise, without the prior written permission of the copyright owners.

This book is available direct from Archaeopress or from our website www.archaeopress.com

Contents

Acknowledgements

This book is an output of the project '*Re-imagining the use of traditional watercraft in the Aegean Sea for a sustainable environment and economy*', which was led by Dr Katerina Velentza in the University of Helsinki with funding from the Kone Foundation (*Koneen Säätiö*, grant no. 202010870). We would like to thank the Kone Foundation for their financial support, as well as their dedication and enthusiasm in supporting bold multidisciplinary, interdisciplinary and transdisciplinary projects that promote academic research, culture and the arts.

Our thanks also go to Professor Kristin Ilves and the team at the Archaeology Unit of the Department of Cultures of the University of Helsinki for hosting the project from 2021 to 2024. Your support and input have been essential in the development of the project and the creation of this book. We are also deeply grateful to our friends and colleagues Dr Veronica Walker Vadillo and Dr Emilia Mataix Ferrándiz for their encouragement in creating this book and inspiration from their thought-provoking interdisciplinary edited volumes. Thank you for all your support in Helsinki.

We would also like to express our gratitude to all the colleagues who peer reviewed the chapters of this book, providing crucial constructive feedback to turn a series of individual papers into this cohesive edited volume that presents cutting-edge maritime archaeological projects during and for the climate emergency that we are all currently experiencing. Your expertise and contribution have been invaluable. This book would not be possible without the assistance of the publishing team in Archaeopress. Thank you, for being extremely supportive and patient with the long process of collating an edited volume.

A big thanks also goes to our current work colleagues, for Katerina at the Energy and Environment Institute of the University of Hull, and for Brandon the Historic Environment Team at Shropshire Council. Our daily work and conversations have shaped our ideas and influenced the creation of this book.

Finally, we are grateful to all our wonderful authors for their hard work. Thank you for your amazing projects and passion in addressing climate change through your archaeological research!

The book editors,

Katerina and Brandon

List of Figures

Chapter 1

Chapter 2

Chapter 3

Chapter 4

Chapter 5

Chapter 6

Chapter 7

Chapter 8

List of Tables

Authors

Daniela Acosta Romero

Anthropologist and final year student in archaeology at the Externado University of Colombia. She has experience in archaeological and historical research, as well as processes related to the management and protection of cultural heritage, cultural resource management, and community initiatives. Daniela has participated in international courses and projects associated with maritime and underwater cultural heritage.

Jesús Aldana Mendoza

Archaeologist specialising in submerged cultural heritage and underwater archaeology, with experience in research, management, and dissemination of maritime heritage in Colombia. His work focuses on the protection of submerged archaeological contexts, including the development of protocols, safeguarding strategies, and management tools, with growing attention to climate change impacts on maritime and coastal cultural landscapes. He has participated in interdisciplinary projects with national cultural and maritime authorities, contributing to studies of maritime cultural landscapes in the Caribbean. He has published and presented research internationally and has professional diving training (PADI Open Water and Advanced Open Water), supporting fieldwork and climate-related monitoring of submerged heritage.

L.Victoria Báez Santos

Archaeologist with a Master's degree in Techniques, Heritage, and Industrial Territories (TPTI) from Paris 1 Panthéon-Sorbonne University, the University of Padua, and the University of Évora (Erasmus Mundus scholar). Her professional experience focuses on research and management of archaeological and maritime heritage in Colombia, with a strong emphasis on the impacts of climate change on submerged and coastal cultural heritage. She has developed methodologies for documentation, risk assessment, and climate monitoring, integrating digital tools such as photogrammetry for heritage protection. She currently works at the Colombian Institute of Anthropology and History in the field of applied heritage technologies and submerged cultural heritage.

Brandon Braun

Brandon has a PhD in Archaeology from the Cotsen Institute of Archaeology at the University of California, Los Angeles (UCLA). He has worked in academic and development-funded archaeology in Greece, Estonia and the United Kingdom, and is currently a county archaeologist providing expert advice for Shropshire Council in the UK. Brandon's primary research interest is commemoration in landscapes, including maritime cultural landscapes. He is also interested in the impact of climate change on cultural heritage, as well as the potential of heritage-led approaches to support climate adaptations, mitigations, and resilience.

Fernando Cadena Duque

Rescue Diver and Master Diver of the Colombian Navy. Fernando also works as a recreational and technical diving instructor certified by PADI, TDI, SDI, and DAN, with specialisations in deep diving, decompression diving, diving in marine protected environments, and mixed gases. His work includes coordinating maritime operations at the Rosario and San Bernardo Coral National Natural Park, conducting underwater environmental conservation and research tasks. He is also involved in the design, planning, and installation of underwater signalling elements, as well as research surveys in archaeological sites and areas dedicated to the conservation of submerged cultural heritage.

Gabriela Caro León

Social sciences and cultural heritage professional with five years of experience in research, historical preservation, and participatory archaeology. She has a background in industrial and contemporary heritage in Cartagena de Indias and Bogotá D.C., and holds certifications in Photogrammetry (generation and quality control of cartographic products), GIS Applications for Archaeology with an environmental focus, and Audiovisual Heritage Management. She has contributed to projects on climate change, urban development, and the evolution of industrialisation processes, as well as integrating academic research with community engagement. She is skilled in project coordination, interdisciplinary research, and collaborative teamwork, and currently works as a researcher in museum, urban, and heritage projects.

Carlos Del Cairo Hurtado

Anthropologist and maritime archaeologist with extensive experience in submerged and coastal cultural heritage research, management, and training. His work focuses on maritime cultural landscapes in Colombia, with increasing emphasis on the impacts of climate change on coastal fortifications, shipwrecks, and traditional maritime practices. He has collaborated with national and international institutions, including UNESCO and the Colombian Ocean Commission, contributing to policy, capacity building, and heritage governance. As a professor and graduate coordinator, he promotes climate-aware approaches to maritime archaeology, integrating traditional navigation knowledge, underwater research, and risk assessment to support the long-term protection of vulnerable coastal and submerged heritage.

Sirine Saad ElDine Ghiye

Sirine is an archaeologist at the General Directorate in Antiquity-Lebanon since 2019 working in maritime missions in Lebanon and abroad. In 2023 she's graduated as Diploma student from the Ulster University programme in Remote Sensing and GIS. Her research focuses on studying climate change using different type of satellite and GIS tools in order to monitor and understand how it impacts maritime archaeological sites.

Deanna Groom

Deanna Groom is a maritime archaeologist who returned, as a mature student, to undertake a programme of PhD research at the School of Ocean Sciences, Bangor University in 2020. Her previous work experience includes undertaking historical research to populate shipwreck and aircraft downed at sea records within the maritime sites and monuments records of Scotland and Wales. She has also worked on the Unpath'd Waters Project at Bangor University (e.g., SS Florrieston https://storymaps.arcgis.com/stories/256a6c01665248e796be6a66b1023128) and undertook an update of the Guide to Good Practice in Marine Survey for the Archaeological Data Service (ADS https://guides.archaeologydataservice.ac.uk/marine/).

José Manuel Matés Luque

Luque, as he is widely known, is a maritime archaeologist and, currently, PhD candidate at the University of Basque Country focusing on maritime and intertidal archaeology in the Basque Country. He has directed some projects like the opening of the Deusto Canal or the landing of transatlantic submarine cables in Bizkaia (Spain). He is a NAS tutor and a freelance archaeologist at www.arqueocean.com offering consultant services for civil engineering projects both on land and in the maritime environment.

Jacob Moe

Jacob Moe is a documentarian and heritage professional with a BA in Politics (Pomona College) and an MFA in Social Documentation (UC Santa Cruz). He is Co-founder of the Syros International Film Festival, as well as Founder and Director of Archipelago Network, an initiative for research and documentation of material knowledge and audiovisual culture in the Cyclades islands. His research practice involves community media, radio and film documentary work, with a focus on peer-to-peer methodologies for knowledge building in island communities.

Carla Riera Andreu

Conservator-restorer of cultural heritage specialising in underwater and maritime archaeology, with a strong focus on *in situ* conservation under changing environmental conditions. Her research addresses the effects of climate change (such as corrosion, sediment dynamics, and biological growth) on submerged archaeological materials, particularly metals and organic remains. She has extensive international training in underwater conservation through the Nautical Archaeology Society. As a lecturer, researcher, and academic coordinator, she leads projects on the protection, monitoring, and sustainable management of coastal and underwater cultural heritage in collaboration with national heritage institutions, integrating climate adaptation strategies into conservation practice.

Bella Romain

Bella is a heritage specialist and consultant based in South Wales. Working across the UK, Bella works directly with professionals and building custodians making decisions on built heritage. Working and living alongside the industrial heritage of Wales, Bella has seen many assets decline due to reducing funding opportunities. For her MSc dissertation, Bella worked to develop a tool to support decision making for heritage at risk, with a focus on coastal heritage under threat due to climate change.

Liliana Rozo Pinzón

Archaeologist from the Externado University of Colombia, with a diploma in Oceanopolitics from the *Escuela Superior de Guerra - Universidad Militar Nueva Granada*. She has completed multiple courses related to underwater cultural heritage and has participated in archaeological projects in both terrestrial and underwater contexts. She is interested in the management and protection of archaeological sites and cultural heritage through the implementation of new technologies, cultural policies, and strategic planning processes framed by community participation, capacity-building, development, and sustainability.

Juan David Sarmiento Rodríguez

Anthropologist, archaeologist, and photographer, researcher at the *Fundación Colombia Anfibia* and at the Colombian Institute of Anthropology and History in the Area of Technologies Applied to Heritage and Underwater Cultural Heritage. He has worked on multiple initiatives related to the research of tangible and intangible heritage associated with shipbuilding, traditional navigation and fishing practices, ethnoarchaeology, and social anthropology, in collaboration with Indigenous, Raizal, Afro-descendant, fishing, and urban communities.

Katerina Velentza

Katerina is a maritime archaeologist and heritage professional with a PhD in Archaeology from the Centre for Maritime Archaeology of the University of Southampton. She is currently a Postdoctoral Research Associate in Environmental Humanities at the University of Hull, in the UK, contributing with heritage perspectives in the research, teaching and PhD supervision of the Energy and Environment Institute. Katerina's research focuses on the interrelationships between archaeology, heritage, climate change and sustainability with project applications in a variety of watery environments and regions, including the Mediterranean, Britain and the Baltic.

Introduction

Katerina Velentza, L. Victoria Báez Santos and Brandon Braun

Maritime archaeology during the climate emergency

The field of maritime archaeology is concerned with the scientific recording, study, and interpretation of the past through the material culture and remains of human activities located in and around aquatic environments including seas, oceans, lakes, and rivers (Bass 2011; Ford *et al.* 2011; Westerdahl 2011: 754). Heritage management is an integral part of the discipline, which has deep-rooted associations with policy interventions in countries around the world. The 2001 Convention on the Protection of the Underwater Cultural Heritage by UNESCO established a clear framework for best practice (Maarleveld *et al.* 2013), bringing together practitioners from around the globe (e.g., Blue and Breen 2019; Trakadas 2012), and shaping legislation in dozens of states around the world (e.g., Aznar 2014; Sharfman *et al.* 2016).

In recent years, faced with the threat of climate change, maritime archaeology has taken up the challenge to act to protect the historic environment and cultural heritage assets in the maritime spectrum, while also contributing to a more just and sustainable present and future (e.g., Gregory *et al.* 2022; Henderson 2019; Henderson *et al.* 2025; Holly *et al.* 2025; Perez-Alvaro 2016; Perez-Alvaro 2025; Rey da Silva 2020: 110; Trakadas *et al.* 2019; Velentza 2023; Wright 2016). Initiatives, such as 2019's Ocean Decade Heritage Network (ODHN n.d.), acknowledge the relevance of maritime archaeology to the implementation of the Ocean Decade (UN Decade of Ocean Science for Sustainable Development, 2021–2030), and interdisciplinary projects, such as CHERISH (2025), highlight the possibility of full integration of climate and heritage research in local policies with simultaneous nature-culture benefits.

Despite being more than halfway through the so-called Decade of Action in the push to deliver the UN Sustainable Development Goals, global policies have struggled to mobilise with sufficient speed to deal with climate change, resulting in very pessimistic climate predictions. In the last two years fossil fuel emissions have increased to an all-time high, and the most recent published reports indicate the year 2024 was the warmest on record globally, passing the average global temperature of 1.6°C above the pre-industrial times (Copernicus Climate Change Service 2025).Despite this stark data, current climate policies have us on track for approximately 2.7°C peak warming by 2100 (Climate Action Tracker 2025; Ripple *et al.* 2024; Romanello *et al.* 2024).

Heritage assets in and around aquatic environments all around the world have already been impacted by climate change effects. Coastal erosion, extreme storminess, storm surges, tidal, fluvial and pluvial flooding, sea-level rise, sea-temperature increase, ocean acidification, frequently in combination with droughts and rising temperatures, and other climate change effects threaten and damage archaeological sites, historic landscapes and buildings. Concurrently, the decline of traditional maritime livelihoods, such as fishing and shipbuilding, and the rise of unsustainable tourism, along with the loss of associated local knowledge within communities and their empirical understanding of maritime regions, have isolated people from their surrounding natural environment and increased their vulnerability to environmental and socio-economic shocks (e.g., Buchan *et al.* 2024; Gillmer 1973; Nadel-Klein 2003). These phenomena are only expected to intensify in the following years. Climate change is here, but heritage practitioners, stakeholders, heritage management organisations and communities are not necessarily prepared or have adequate resources to deal with these circumstances, especially due to the financial recession that most sectors are experiencing.

The future by the sea

This edited volume features an array of maritime heritage projects that have taken action to record, monitor and protect maritime heritage in the face of the climate emergency. The authors first came together during the online symposium 'The Future by the Sea: Engaging with maritime archaeological research during the climate emergency' in November 2023. It was organised by Katerina Velentza at the University of Helsinki as part of the project *Re-imagining the use of traditional watercraft in the Aegean Sea for a sustainable environment and economy 2021-2024* (n.d.) with the help of L. Victoria Báez Santos from the NGO *Fundación Colombia Anfibia*. The aim of the event was to bring together early career researchers working on innovative projects that link maritime archaeology and climate change around the world. The idea of creating this edited volume came from the lively discussions and the converging points of the projects presented in that event, along with the obvious need to promote the work of early career scholars who, despite their passion and knowledge of the subject, are frequently obstructed from participating in academic and sector-wide discussions, whether due to shrinking job markets or limited availability of resources in the sectors of heritage and archaeology.

This volume presents eight chapters that showcase completed or ongoing projects of early career researchers and practitioners from distinct parts of the world. The projects illustrate how the discipline of maritime archaeology can contribute to climate action, sustainability, adaptation and climate resilience, while also helping to mitigate some of the main risks that heritage linked to aquatic environments is experiencing and will experience due to the climate breakdown. The researchers have not simply recorded tangible heritage assets and archaeological sites that might be impacted by climate change effects, but rather, their work involves understanding, documenting or modelling climate change impacts in their local regions, reviewing local policies and making recommendations, and working closely with local stakeholders and communities to find sustainable solutions that best safeguard tangible and intangible heritage assets. Above all, though, they are determined to preserve heritage for future generations despite the climate emergency. While recognising the limitations and challenges brought up by climate change and its socio-economic implications, as well as the defunding of the archaeology and heritage sectors, the projects presented in this book also show how talented early career scholars and practitioners can make substantial contributions to communities through volunteering work, small grants, and project-specific academic funding. Overall, each chapter of this book proves how small-scale, local initiatives by passionate people can make a large impact.

The volume starts with Chapter 1 and the results of a research programme that explores the impacts of climate change on historic shipwrecks under water and in the intertidal zone of Wales, in the United Kingdom. This interdisciplinary study identifies various climate change impacts that will affect the longevity of many underwater and intertidal cultural heritage assets, while also highlighting the significance of international cross-disciplinary studies combined with stakeholder collaboration to tackle known threats. On a similar note, Chapter 2 presents a variety of climate change effects that are already affecting intertidal archaeological sites in the Basque Country, in Spain. Sea level rise, coastal erosion, changes in water temperatures and pH levels, as well as the presence or absence of different organisms were observed in river, estuarine and coastal shipwrecks in the area. This data can offer unique local climate data and place-based observations on the impacts of climate change on various types of material culture in the region.

In Chapter 3, the focus of the book shifts to methodological approaches that could be used to track and predict impacts of climate change on coastal archaeological sites. Through a case study from northern Lebanon, this project uses data from satellite imagery, including associated measurements of Sea Surface Temperature (SST) and Sea Surface Salinity (SSS), as well as observations and mapping of coastal changes over long periods of time, to track noticeable impacts of climate change, and also model potential repercussions and future vulnerabilities for coastal heritage assets. Subsequently, Chapter 4 presents another innovative methodological approach, the use of Multi-Attribute Value Theory (MAVT) that could be applied to conservation management strategies for heritage at risk. The case study of Whitford Point Lighthouse in Gower, Swansea, Wales, UK, demonstrates how MAVT could be used in practice to support decision making for managing complex heritage assets during the climate emergency.

Chapter 5 introduces the work carried out by the NGO *Fundación Colombia Anfibia* in Colombia. In this paper, the transdisciplinary methodology used in Cartagena de Indias to assess the effects of climate change on the city's tangible and intangible heritage is presented. As the preliminary results of the project show, this methodology has created a low-cost transdisciplinary and co-created understanding of climate change in Cartagena de Indias and initiated contact and dialogue between significant members of the city. Following up on these observations, Chapter 6 presents in more detail aspects of the project *Colaborartorio Azul* of the *Fundación Colombia Anfibia.* Through a case study focusing on the island of Tierrabomba (*Isla de Tierrabomba*) in Cartagena de Indias, the approaches used to instigate cooperation, co-production and active participation of community members and stakeholders are explained. The co-production and participatory research of the project have been enhancing the community's decision-making capacity and increasing opportunities for knowledge exchange through public mechanisms that address the current global climate emergency.

In Chapter 7 intangible maritime heritage is brought to the fore. A summary of the observations from the project *Re-imagining the use of traditional watercraft in the Aegean Sea for a sustainable environment and economy* highlights the potential of maritime heritage to assist in restoring the balance between humans and nature in the dynamic environment of the eastern Mediterranean. Historical, archaeological and ethnographic approaches can assist communities to build environmental awareness and realise sustainability opportunities from their past, their culture and tradition that could help populations adapt in the face of the current climate emergency and environmental deterioration. Chapter 8 presents the project *Cycladic Maritime Trades*, carried out also in the Aegean Sea, Greece, which proposes maritime heritage documentation as a catalyst for pursuing future sustainable development in island communities facing accelerating social and environmental change in the first quarter of the 21st century.

Throughout the eight chapters of the book, the authors underpin the need for better dialogue between heritage, environmental and climate change practitioners, communities and policymakers at a local, regional and international level to simultaneously preserve natural and cultural heritage, as described in the United Nations Sustainable Development Goal Target 11.4. The projects presented in this book from the UK, Spain, Lebanon, Colombia and Greece all follow interdisciplinary or transdisciplinary perspectives in an attempt to bridge policy and practice, while forging meaningful collaborations between communities, NGOs, academics and stakeholders. All projects highlight how, as archaeologists and heritage professionals, we possess significant tools and skills that can assist in assessing risk, responding to heritage loss, but also initiating actionable recommendations and policy advice that could have an impact on both a local and global scale. We hope that our work presented in this edited collection will inspire and encourage more climate-change-related actions in the field of archaeology and heritage.

References

Aznar, M.J. 2014. The Contiguous Zone as an Archaeological Maritime Zone. *International Journal of Marine and Coastal Law* 29: 1–51.

Bass, G. 2011. The Development of Maritime Archaeology, in B. Ford, D.L. Hamilton and A. Catsambis (eds) *The Oxford Handbook of Maritime Archaeology*: 3–24. Oxford: Oxford University Press. <https://doi.org/10.1093/oxfordhb/9780199336005.013.0000>

Blue, L. and C. Breen. 2019. Maritime Archaeology and Capacity Development in the Global South. *Journal of Maritime Archaeology* 14: 321–332. <https://doi.org/10.1007/s11457-019-09244-x>

Buchan, P.M., L.D. Glithero, E. McKinley, M. Strand, G. Champion, S. Kochalski, K. Velentza, R.A. Praptiwi, J. Jung, M.C. Márquez, M.V. Marra, L.M. Abels, A.L. Neilson, J. Spavieri, K.E. Whittey, M.M. Samuel, R. Hale, A. Čermák, D. Whyte, L. West, M. Stithou, T.J. Hegland, E.S. Morris-Webb, V. Flander-Putrle, P. Schiefer, S. Sutton, C. Onwubiko, O. Adeoye, A. Akpan and D.L. Payne 2024. A transdisciplinary co-conceptualisation of marine identity. *People And Nature*. <https://doi.org/10.1002/pan3.10715>

CHERISH (Climate Change and Coastal Heritage), viewed 10 September 2025, <https://cherishproject.eu/en/>.

Climate Action Tracker, The CAT Thermometer, November 2025 viewed 10 December 2025, <https://climateactiontracker.org/global/cat-thermometer/>.

Copernicus Climate Change Service, The 2024 Annual Climate Summary, Global Climate Highlights 2024, viewed 10 September 2025, <https://climate.copernicus.eu/global-climate-highlights-2024>.

Ford, B., D. Hamilton and A. Catsambis 2011. *The Oxford Handbook of Maritime Archaeology*. Oxford: Oxford University Press. <https://doi.org/10.1093/oxfordhb/9780199336005.001.0001>

Gillmer, T. 1973. *Working Watercraft: A Survey of the Surviving Local Boats of Europe and America*. London: Patrick Stephens.

Gregory, D., T. Dawson, D. Elkin, H. Van Tilburg, C. Underwood, V. Richards, A. Viduka, K. Westley, J. Wright and J. Hollesen 2022. Of time and tide: the complex impacts of climate change on coastal and underwater cultural heritage. *Antiquity* 96.390: 1396–1411. <https://doi.org/10.15184/aqy.2022.115>

Henderson, J. 2019. Oceans without History? Marine Cultural Heritage and the Sustainable Development Agenda. *Sustainability* 11.18: 5080. <https://doi.org/10.3390/su11185080>

Henderson, J., G. Holly, A. Rey da Silva and A. Trakadas 2025. The Cultural Heritage Framework Programme: Highlighting the contribution of Marine Cultural Heritage to the UN decade of Ocean Science for sustainable development (2021–2030). *Oceans* 6 1: 1-19. <https://doi.org/10.3390/oceans6010001>

Holly, G., J. Henderson, A. Edwards and H. Cocks 2025. Marine Cultural Heritage as a Catalyst for Sustainable Ocean Practices. One Ocean Science Congress 2025, Nice, France, 3–6 Jun 2025, OOS2025-981. <https://doi.org/10.5194/oos2025-981>

Maarleveld, T., U. Guérin and B. Egger 2013. *Manual for Activities directed at Underwater Cultural Heritage: Guidelines to the Annex of the UNESCO 2001 Convention*. Paris: United Nations Educational, Scientific and Cultural Organization.

Nadel-Klein, J. 2003. *Fishing for heritage: Modernity and loss along the Scottish coast*. Abingdon: Routledge.

Ocean Decade Heritage Network – A Global Initiative, viewed 10 September 2025, <https://www.oceandecadeheritage.org/>.

Perez-Alvaro, E. 2016. Climate Change and Underwater Cultural Heritage: Impacts and Challenges. *Journal of Cultural Heritage* 21: 842–848. <https://doi.org/10. 1016/j.culher.2016.03.006>

Perez-Alvaro, E. 2025. Preserving the past, shaping the future: the impact of underwater cultural heritage on global development. *Journal of Cultural Heritage Management and Sustainable Development*. <https://doi.org/10.1108/JCHMSD-03-2023-0033>

Re-imagining the Use of Traditional Watercraft in the Aegean Sea for a Sustainable Environment and Economy, viewed 10 September 2025, <https://traditionalwatercraftaegeansea.wordpress.com/>.

Rey da Silva, A. 2020. Sailing the Waters of Sustainability: Reflections on the Future of Maritime Cultural Heritage Protection in the Global Sea of Development. Post- Classical Archaeologies 10: 107–34.

Ripple, W.J., C. Wolf, J.W, Gregg, J. Rockström, M.E. Mann, N. Oreskes, T.M, Lenton, S. Rahmstorf, T.M. Newsome, C. Xu, J.C. Svenning, C. Cardoso Pereira, B.E. Law and T.W. Crowther 2024. The 2024 state of the climate report: Perilous times on planet Earth. *BioScience* 74.12: 812–824. <https://doi.org/10.1093/biosci/biae087>

Romanello M., M. Walawender, S.C. Hsu, A. Moskeland, Y. Palmeiro-Silva, D. Scamman, Z. Ali, N. Ameli, D. Angelova, S. Ayeb-Karlsson, S. Basart, J. Beagley. P.J. Beggs, L. Blanco-Villafuerte, W. Cai, M. Callaghan, D. Campbell-Lendrum, J.D. Chambers, V. Chicmana-Zapata, L. Chu, T.J. Cross, K.R. van Daalen, C. Dalin, N. Dasandi, S. Dasgupta, M. Davies, R. Dubrow, M.J. Eckelman, J.D. Ford, C. Freyberg, O. Gasparyan, G. Gordon-Strachan, M. Grubb, S.H. Gunther, I. Hamilton, Y. Hang, R. Hänninen, S. Hartinger, K. He, J. Heidecke, J.J. Hess, L. Jamart, S. Jankin, H. Jatkar, O. Jay, I. Kelman, H. Kennard, G. Kiesewetter, P. Kinney, D. Kniveton, R. Kouznetsov, P. Lampard, J.K.W. Lee, B. Lemke, B. Li, Y. Liu, Z. Liu , A. Llabrés-Brustenga, M. Lott, R. Lowe, J. Martinez-Urtaza, M. Maslin, L. McAllister, C. McMichael, Z. Mi, J. Milner, K. Minor, J. Minx, N. Mohajeri, N.C. Momen, M. Moradi-Lakeh, K. Morrisey, S. Munzert, K.A. Murray, N. Obradovich, M.B. O'Hare, C. Oliveira, T. Oreszczyn, M. Otto, F. Owfi, O.L. Pearman, F. Pega, A.J. Perishing, A.C. Pinho-Gomes, J. Ponmattam, M. Rabbaniha, J. Rickman, E. Robinson, J. Rocklöv, D. Rojas-Rueda, R.N. Salas, J.C. Semenza, J.D. Sherman, J. Shumake-Guillemot, P. Singh, H. Sjödin, J. Slater, M. Sofiev, C. Sorensen, M. Springmann, Z. Stalhandske, J.D. Stowell, M. Tabatabaei, J. Taylor, D. Tong, C. Tonne, M. Treskova, J.A. Trinanes, A. Uppstu, F.Wagner, L. Warnecke, H. Whitcombe, P. Xian, C. Zavaleta-Cortijo, C. Zhang ,R. Zhang, S. Zhang, Y. Zhang, Q. Zhu, P. Gong, H. Montgomery and A. Costello 2024. The 2024 report of the Lancet Countdown on health and climate change: facing record-breaking threats from delayed action. *Lancet* 404.10465: 1847–1896.

Sharfman, J., J. Boshoff and J. Gribble 2017. Benefits, Burdens, and Opportunities in South Africa: The Implications of Ratifying the 2001 UNESCO Convention on the Protection of Underwater Cultural Heritage, in L. Harris (ed.) *Sea Ports and Sea Power*. Cham: Springer. <https://doi.org/10.1007/978-3-319-46985-0_9>

Trakadas, A. 2012. Maritime Archaeology and Mitigation in Africa: Endangered Resources, Policies, and Practice. *Journal of Maritime Archaeology* 7: 3–8. <https://doi.org/10.1007/s11457-012-9100-2>

Trakadas, A., A. Firth, D. Gregory, D. Elkin, U. Guerin, J. Henderson, J. Kimura, D. Scott-Ireton, Y. Shashoua, C. Underwood and A. Viduka. 2019. The Ocean Decade Heritage Network: Integrating Cultural Heritage Within the UN Decade of Ocean Science 2021–2030. *Journal of Maritime Archaeology* 14: 153–165. <https://doi.org/10.1007/s11457-019-09241-0>

Velentza, K. 2023. Maritime Archaeological Research, Sustainability, and Climate Resilience. *European Journal of Archaeology* 26.3: 359–377. <https://doi.org/10.1017/eaa.2022.48.>

Westerdahl, C. 2011 The Maritime Cultural Landscape, in B. Ford, D.L. Hamilton and A. Catsambis (eds) *The Oxford Handbook of Maritime Archaeology*: 735–762. Oxford: Oxford University Press. <https://doi.org/10.1093/oxfordhb/9780199336005.013.0032>

Wright, J. 2016. Maritime Archaeology and Climate Change: An Invitation. *Journal of Maritime Archaeology* 11: 255–270. <https://doi.org/10.1007/s11457-016-9164-5>

1.
Historic Shipwrecks and the Impacts of Climate Change

Deanna Groom

This paper presents the results of a four-year research programme exploring the impacts of climate change on historic shipwrecks under water and in the intertidal zone of Wales. With the use of interdisciplinary methodology, the project identifies that many historic shipwreck sites frequently fall within the Welsh storm wave base zone and will be at risk of sediment movements more frequently and more severely by 2100. Sites in the intertidal zone will be subjected to increased physical force and will experience shorter periods of calm weather during which sediment is returned onshore to re-cover sites. Several warmer water species are confirmed as expanding their range northwards in response to warming seas, while at the same time the biodiversity of Welsh shipwrecks has decreased by 7%. All the results correspond with wider trends reported for the natural marine environment of Wales in relation to climate change. In addition to the above observations, the project undertook a worldwide practitioners' survey to identify research priorities that could protect underwater and intertidal heritage assets. The survey also explored the possibility of creating a network of 'Climate Change Indicator Wrecks', which would allow research of specific case studies and climate change impacts in more depth through international collaboration. The concept of a 'Network of Climate Change Indicator Wrecks' or 'Sentinel Sites' has already begun in Wales through the creation of baseline surveys for 17 intertidal sites using photogrammetry to record the changes in the material remains as they happen in real time while climate change phenomena intensify. At the same time environmental assessments through ecological surveys for a further 52 permanently submerged sites help identify species with the potential to act as biomarkers for the impact of warming seas on underwater material remains. This study makes an important contribution, at a nation-wide scale, in reviewing the risks of the changes that are already taking place at shipwreck sites during the most critical decade of the climate emergency. Additionally, it considers the potential of research and policy actions at an international scale that could ensure more up-to-date knowledge and safeguarding guidelines of underwater and intertidal shipwrecks.

Introduction

Shipwrecks contain a vast and diverse reserve of evidence for our seafaring past and also provide unique biodiversity hotspots for grazers, predators, scavengers and detritivores amongst a rich turf of sponges, tunicates, cnidarians, and bryozoans. The role of shipwrecks in sustaining juvenile fish populations is also well recognised in existing research (e.g., Paxton *et al.* 2024). The significance of in-situ preservation and management of shipwrecks, with non-destructive and non-intrusive conservation strategies, is emphasised in both the ICOMOS Charter on the Protection and Management of Underwater Cultural Heritage 1996 (ICOMOS 1997: Article 1) and the 2001 UNESCO Convention on the Protection of Underwater Cultural Heritage (UNESCO 2001: Article 2) as key to best safeguard the cultural and natural heritage on the seabed. Excavation and removal of the assets is perceived as necessary only when sites are at risk due to circumstances which cannot be mitigated (Manders 2008: 34).

In recent years, heritage organizations and archaeological practitioners have begun to recognise that the marine environmental parameters, which have so far been assessed to assist *in-situ* preservation of shipwrecks, are slowly and relentlessly changing due to climate change. Warming seas, increased storminess, ocean acidification and increasing numbers of hypoxia events, as well as changes in marine biodiversity and sediment flux chemical exchange processes are bringing new threats to underwater and intertidal heritage (Figure 1.1), which need to be assessed, understood and quantified, especially due to the rhythm in which the climate emergency is intensifying (Dunkley 2015; Gearheart *et al.* 2011; Kintisch 2016; Mustow 2021; Stieglitz and Waterson 2013). Researchers and heritage stakeholders have been attempting to document these changes and act for the better safeguarding of shipwrecks at risk. For instance, in the UK, an article published in 2009 by members of Historic England's coastal and marine team assessed different hulk vessels situated along the UK coast in an attempt of making suggestions as to how to best manage the inevitable disintegration and loss of such sites during the climate emergency (Murphy *et al.* 2009). The article identified four main categories of impact including: direct physical impacts causing accelerated erosion or increased flooding; indirect impacts which are the consequence of decisions taken now by coastal managers; indirect impacts related to offsetting climate change through development (such as the expansion of the renewable energy sector and the wider development of offshore windfarms); and the northwards expansion of non-native and invasive species, such as the species of shipworm *Lyrodus pedicellatus*, which has already been noted as a special concern in England, with examples recorded off Cornwall, in Langstone Harbour in Hampshire, and on the Mary Rose wreck in the Solent (Murphy *et al.* 2008).

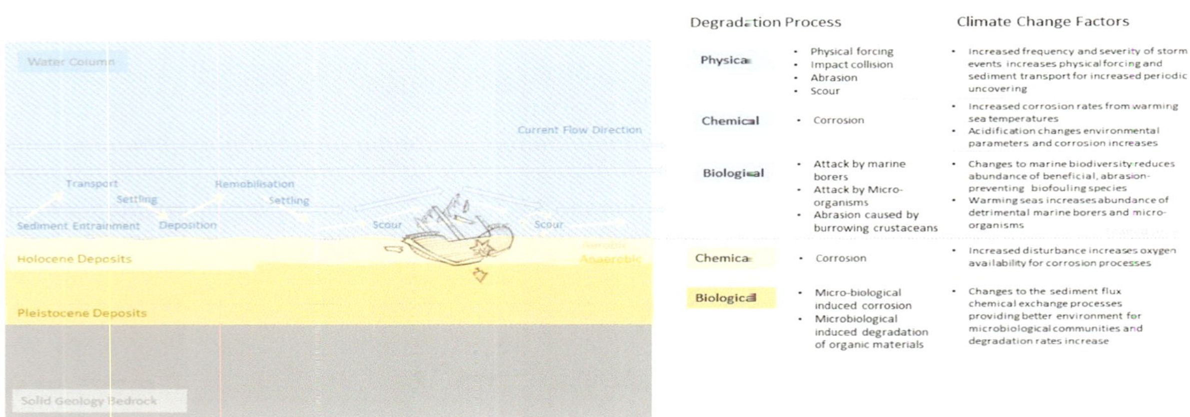

Figure 1.1: Processes of degradation and where they take place in the water column and seabed. Diagram made after Gjelstrup Bjordal et al. 2011: 109. Copyright: Deanna Groom.

Other academic studies have identified similar risks to shipwreck sites from climate change effects around the world. For example, Perez-Alvaro (2016) in her review paper notes four main climate-related changes likely to affect underwater heritage, including: warming oceans; changes to ocean currents; chemical changes; and sea level rise. Closely associated to the risks above are threats to underwater cultural heritage for human response to the climate crisis, such as the exploitation of new fishing grounds for bottom trawling, as fish species change their range in response to warming seas (e.g., Jarvis 2024). Similarly, new oil and gas reserves, offshore renewables developments, and the exploration of deep-sea mining opportunities, can be linked back to countries failing to meet the global warming targets of the 2015 Paris Agreement (Jarvis *et al* 2023).

Through a review of studies such as the above, it became evident that shipwreck research in the 21st century needs to move beyond simply modelling site formation processes and add avenues to properly incorporate climate change in the site assessments. My research project, undertaken at the School of Ocean Sciences, Bangor University from 2021 to 2024, has taken a first step towards addressing these concerns for the shipwreck sites of Wales. The preliminary results and observations of this study are presented in the following sections in the hope that it will inspire others to take a closer look at site data they have collated over the years and develop new investigation practices that will truly inform the future preservation of underwater cultural heritage sites.

Methodology

The geographical region of the study comprises part of the spatial area defined in the Welsh National Marine Plan, which covers 2120km of coastline and an offshore area of 32,000km^2. Within this area, there are the remains of nearly 1000 known shipwrecks and aircraft downed at sea. Nearly a quarter of these known sites are from the First and Second World Wars. Furthermore, there is the potential for an additional 4500 sites to be present according to historical documentary sources, such as newspaper accounts. These heritage assets represent the cultural property of 32 nations for which the Welsh Government takes responsibility to manage and conserve (Figure 1.2).

The first stage of the project involved carrying out a practitioners' survey. This involved an online questionnaire, which was circulated to archaeologists and heritage professionals worldwide through existing research networks. The objective was to obtain an understanding of the various challenges being experienced in shipwreck site management during the climate emergency. Participants were offered the opportunity to suggest the climate change parameters they felt to be most impactful and identify research priorities, hence gauging the potential to develop more universal climate change impact assessment criteria which could be used worldwide. 27 practitioners took part in this survey from 11 countries including Wales, England, Northern Ireland, Norway, Finland, Belgium, Malta, the United States of America, Tanzania, Australia, and Japan (Figure 1.3).

In the second stage of the project, an extensive literature review of oceanic climate change predictions and archaeological site formation processes was undertaken. A series of matrices were developed to assess climate change effects (e.g., increased storminess, warming seas and ocean acidification) and their likely impact on known physical, biological and chemical site formation processes of heritage sites. An ArcGIS geodatabase was developed to collate datasets that underpin an understanding of known physical, biological and chemical processes, such as salinity, temperature, acidity, wave climate, tidal flows, nature of sediment and thresholds of motion as presently observed. A sampling grid was created for the offshore of Welsh waters and was used to create a shapefile which collated information about present conditions and projected changes to the year 2050. Significant wave height datasets were also assembled from four semi-permanent wave buoys around the Welsh coast from which was calculated average depths of storm wave base in the previous 20 years (i.e., wave base being the depth beneath the sea surface at which a passing wave begins to set sediment transport in motion). A shapefile was created to suggest the most common present extents of storm wave base zone.

Figure 1.2: Study area encompassing area cover by the Welsh National Marine Plan with known and documented losses (Welsh National Marine Plan | GOV.WALES). Copyright: Deanna Groom.

Figure 1.3: Countries where practitioners responded to the questionnaire. Copyright: Deanna Groom.

Additionally, coastal vulnerability was explored through the development of another ArcGIS shapefile to collate information about the coast edge, such as nature of sediment supply, cliff material and erosion rates, sediment transport directions from longshore drift at beach level, and beaches with a known history of draw down and rebuild on an annual basis. This resulted in identifying priority stretches of coastline to focus monitoring activity, when correlated with known intertidal wrecks sites. To complement the desk-based work done in GIS, 32 reconnaissance field visits were undertaken, from which 17 sites were selected and taken forward into the three-year monitoring programme of the project (Table 1.1). For these sites, photogrammetric surveys were used to record patterns of site uncovering, in response to seasonal changes in beach profiles. When wave energy is high, such as through winter periods, sediment is drawn offshore. When calmer weather prevails, sediment is brought back by the asymmetry between sediment transport by waves running up the shore and the sediment transport that prevails in backwash. The photogrammetric surveys were combined with the information contained in beach profiles recorded for local government authorities from 2019 onwards held by the Welsh Coastal Monitoring Centre. Although these profiles have been primarily used to inform coastal engineering projects, they are now proving exceptionally useful for monitoring sediment changes in the vicinity of intertidal wrecks.

Finally, the Marine Conservation Society supplied ecological surveys for 52 underwater wreck sites from their 'Seasearch' initiative. The surveys spanned the years 1995-2013. These were collated into a Microsoft Access database, which was enhanced with information about abundance, preferred environmental conditions, and distinctions between local and invasive non-native species.

Table 1.1: Intertidal Wreck sites taken in the monitoring programme. Copyright: Deanna Groom.

HSICC Site Code	National Primary Resource Number	Name	Vessel Type	Latitude WGS84	Longitude WGS84	Monitoring	Potential to act as a Climate Change Indicator Wreck
AB01	273277	Progress	Wooden Ketch built in 1863 at Appledore, Devon, abandoned 1954	51.68346	-5.08179	Y	Low
CS05	302164	Unnamed Wreck	Believed to be wooden barque Vittoria, Italian Registry, driven ashore 1872	51.68331	-4.33525	Y	High
CS06	544681	Unnamed Wreck	Possible Ieisides, wooden barque, Italian Registry, driven ashore 1872	51.69162	-4.34719	Y	High
CS08	544616	Unnamed Wreck	Remains of wooden schooner or small brig, history unknown	51.6936	-4.34865	Y	High
CS12	544676	Unnamed Wreck	Believed to be a wooden sailing trow, history unknown	51.69288	-4.34615	Y	High
MAL13	271570	Athena	Wooden brig built 1840, home port Candia (Heraklion) Crete, driven ashore 1852	53.15286	-4.41419	Y	High
MUM19	309544	Unnamed Wreck	Believed to be a wooden fishing vessel laid up possibly pre-World War I	51.57434	-3.99341	Y	High
MUM20	544839	Unnamed Wreck	Small wooden fishing vessel, remains of traditional craft of Mumbles oyster fishing fleet, laid up post-World War I	51.57477	-3.99657	Y	High
MUM21	544842	Unnamed Wreck	Small wooden fishing vessel, remains of traditional craft of Mumbles oyster fishing fleet, laid up post-World War I	51.575	-3.99691	Y	High
MUM22	544840	Unnamed Wreck	Small wooden fishing vessel, remains of traditional craft of Mumbles oyster fishing fleet, laid up post-World War I	51.57537	-3.9959	Y	High

HSICC Site Code	National Primary Resource Number	Name	Vessel Type	Latitude WGS84	Longitude WGS84	Monitoring	Potential to act as a Climate Change Indicator Wreck
MUM23	544841	Unnamed Wreck	Small wooden fishing vessel, remains of traditional craft of Mumbles oyster fishing fleet, laid up post-World War I	51.5758	-3.99646	Y	High
PEN16	1001	Francis Beddoe	Wooden Ketch built 1872 at Saundersfoot, Wales, driven shore 1924	51.74087	-4.54288	Y	High
RB03	273914	Helvetia	Wooden barque built in 1855, home port Bremerhaven, drive ashore 1877	51.57375	-4.29049	Y	High
RB04	273959	Vennerne	Iron hull barque, built in 1864 at Preston, Lancashire, driven ashore 1894	51.56921	-4.29492	Y	High
SHE02	410571	Unnamed Wreck	Believed to be a wooden ketch or small schooner, history unknown	51.73352	-5.10028	Y	Medium
SHW10	544678	Unnamed Wreck	Wooden steam trawler laid up/broken up after fishing industry downturn after World War I	51.72586	-5.10627	Y	Medium
SHW33	544677	Unnamed Wreck	Wooden steam trawler laid up/broken up after fishing industry downturn after World War I	51.72807	-5.10696	Y	Medium
YNS17	407989	Unnamed Wreck	Believed to be one of the last wooden, sailing ferries on the Dyfi estuary laid up in 1864	52.52745	-4.04234	Y	Medium

Results

Practitioners' Survey

The online questionnaire for the practitioners' survey was shared with archaeologists and heritage professionals from April to September 2021. Of the responders, 23% worked in government agencies or national custodian agencies, 53% worked in academia, and 15% were from organisations such as commercial archaeology units, archaeological contractors and not-for-profit research institutes. A further 4% was part of the recreational sports diving community.

The questionnaire included open-ended questions, which allowed participants to provide typed comments, and Likert scales, in which the participants could rank their agreement to various statements on a five-point scale ranging from 'extremely important' to 'not at all important'. Of foremost interest, participants identified the impact of frequent and severe storms as a major threat to shipwreck sites. Other priority areas for research identified by practitioners included (Figure 1.4):

- gaining fuller understanding of the range and scale of sites being impacted;
- gaining a better understanding of changes in shipwreck ecosystems from warming seas;
- gaining a better understanding of the impact of climate change on the interrelationships between the physical, biological, and chemical processes of archaeological site formation;
- expanding our understanding of the impacts of ocean acidification on metal wrecks and corrosion processes;
- gaining a better understanding of the impact of storms on shipwrecks situated in medium (15m-50m) sea depths;
- developing new holistic interdisciplinary approaches which involve marine and social sciences;
- developing a broader standardisation for marine survey data collection for climate change purposes, so that to determine best practice standards that could be shared and applied internationally.

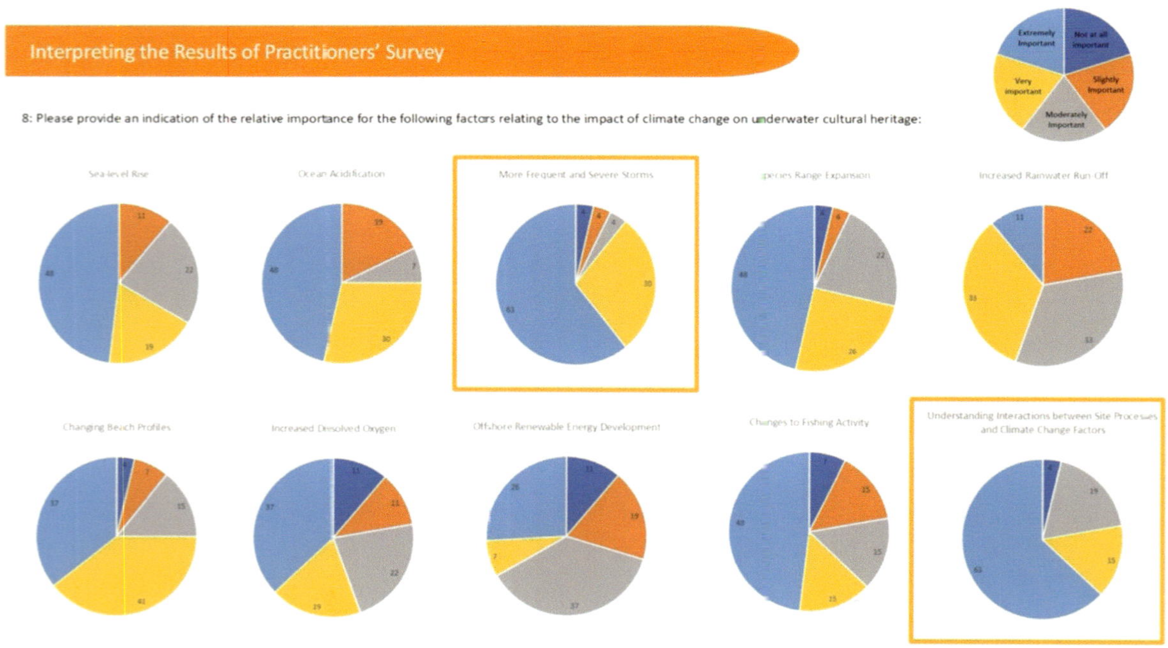

Figure 1.4: Result of practitioner's questionnaire. A five-point Likert scale was used to identify the levels of agreement with statements (i.e., scale of responses ranging from 'extremely important' to 'not important at all'). Copyright: Deanna Groom.

The responses seen in the practitioners' survey guided the scope of the research of the project taken forward. Based on this and the gaps identified through a literature review, the decision was made to focus on three key climate change factors: the impact of more frequent and severe storms, ocean acidification, and warming seas.

Development of Matrices

As part of studying the impacts of more frequent and severe storms, ocean acidification and warming seas, three matrices were formulated to help with the assessment of these phenomena on underwater and intertidal shipwrecks in Wales (Tables 1.2–1.4).

In the creation of the matrix exploring impacts of a severe storm on shipwreck sites (Table 1.2), increased exposure and physical impact from wind and waves, variations in oxygen levels, increased water levels, coastal erosion, shoreline retreat and re-alignment, and increased rainwater run-off and flooding, were all considered for their impact on physical, biological and chemical processes of shipwreck site formation.

The matrix considered climate models of the Couple Model Intercomparison Project 5 (CMIP 5), a plausible extreme scenario of a 50–80% increase in days of strong winds in the United Kingdom by 2070–2100, in comparison to the weather experienced in the period 1975–2005. Another feature that was incorporated in the matrix is the prediction that larger extreme waves will be experienced on Atlantic facing coasts and that there will also be increases in Significant Wave Heights in the Irish Sea as noted by Bricheno and Wolf (2018).

Another matrix (Table 1.3) explored the impacts of increasing sea temperatures on shipwrecks of Wales, a significant phenomenon given the projected increase of 4-6°C for summer (Tinker and Howes 2020: 30) and with stratification beginning seven days earlier and remaining five to ten days longer (Sharples *et al.* 2020; Young and Holt 2007). The matrix regarding the impacts of increased sea temperature explored the increase in corrosion rates, but also the buffering of the impact that might be created by lower dissolved oxygen and an increase in the growth rates of any colonising organisms. This poses questions relating to how much of a buffering effect biofouling may be able to provide if it is caused by hypoxia events. For example, if we look at a metal wreck, underneath the biofilm layer, immediately next to the metal surface, anaerobic conditions are created encouraging bacteria growth. The enzymes secreted by these bacteria exert a major influence on the kinetics of corrosion as well as on the nature of the corrosion products. A difference of four pH units, equating to 1:10,000 greater acidity, has been found inside concretions (Gjelstrup Bjordal *et al.* 2011: 95). Will increasing sea temperatures ocean acidification reduce or increase the abundance of these bacteria acting on the surface of metals?

This query led to the need to explore ocean acidification through a separate matrix (Table 1.4). Through this research it was identified that average carbon dioxide concentration now exceeds 400 parts per million (ppm) in the waters surrounding the Welsh coast. Over the past 10 years, this concentration has increased 2.3ppm annually. Models are predicting a drop in mean pH across the continental shelf of about 0.0036 per year by 2010 and significant spatial variation such as 0.005 per year in Bristol Channel and 0.002 per year in the Celtic Sea (Humphreys *et al.* 2020: 54). This suggests that chemical and biological processes would be most impacted by ocean acidification. However, there is uncertainty regarding CO2 uptake and magnitude of pH change in low and medium salinity waters such as estuaries and near-coast waters (Howarth 2005), where, for the case of Wales, most historic shipwrecks are located.

Table 1.2: Matrix exploring the impact of a severe storm on the physical, chemical, and biological process of shipwreck site formation. Copyright: Deanna Groom.

Input	More frequent storm events and increase in most-severe wave heights								
	Physical Processes Impact			Biological Processes Impact			Chemical Processes Impact		
	Process	Cause	Impact	Process	Cause	Impact	Process	Cause	Impact
Increased wind speed with changes of direction	Increased physical impact on structure	Increased wind and significant wave heights	Negative Impact: Increased physical force on vessel structure						
Increase sediment in suspension	Increased mechanical abrasion	Scour	Negative Impact: Increased abrasion causing loss of surface details	Loss of protective biological cover (fouling biota)	Increased abrasion of structure above seabed	Negative Impact: New surfaces provided for colonisation by marine wood-borers leading to increased degradation	Change in character of overlying sediment changing the chemical composition of the preservation environment	New sediment is deposited over the site	Negative impact: Rates of microbiological and fungal decay increase with a move from finer to coarser deposits Positive Impact: Rates of microbial and fungal decay slow down, with move from coarser to finer deposits
Increased dissolved oxygen in turbulent waters				Rapidly moving and disturbed water (e.g., storm waves, groundwater discharge) contains more dissolved oxygen	Proliferation of marine wood-borers, fungi, and bacteria responsible for degradation	Negative Impact: Changes to aerobic degradation mechanisms cause increased rates of degradation/ corrosion	Rapidly moving and disturbed water (e.g., storm waves, groundwater discharge) contains more dissolved oxygen	Increased galvanic corrosion activity and proliferation of fungi and bacteria responsible for degradation	Negative Impact: Increased corrosion and thinning of metallic remains (e.g., plating)
Beaches change slope/profile	Beaches flatten to absorb more wave energy	Exposure of site due to beach-draw-down or reduction of depth of burial sediment	Negative Impact: Site moves from preservation state in near equilibrium to active degradation with increased wetting and drying	Increased aerobic biological activity	Colonisation of newly exposed material and infestation of marine wood-borers	Negative Impact: Site moves from preservation state in near equilibrium to active degradation	Marine wood-borers and degrading bacteria and fungi are provided with new material for life-cycle consumption	Changes to aerobic and anaerobic mechanisms cause increased rates of degradation/ corrosion rate	Negative Impact: Site moves from preservation state in near equilibrium to active degradation

Input	More frequent storm events and increase in most-severe wave heights								
	Physical Processes Impact			Biological Processes Impact			Chemical Processes Impact		
	Process	Cause	Impact	Process	Cause	Impact	Process	Cause	Impact
Permanent Coastal Squeeze	Nearshore depth profile steepens where hard coastal defences have been created and archaeological remains area exposed for longer periods	Offshore water levels increase in in response to steepening and in combination with sea-level rise	Negative Impact: Site moves from preservation state in near equilibrium to active degradation	Increased aerobic biological activity causing an increase in degradation rates	Increase infestation of marine wood-borers	Site moves from preservation state in near equilibrium to active degradation	Marine wood-borers and degrading bacteria and fungi are provided with new material for life-cycle consumption	Changes to aerobic and anaerobic mechanisms cause increased rates of degradation/ corrosion rate	Negative impact: Site moves from preservation state in near equilibrium to active degradation
Shoreline retreat and re-alignment	Erosion of sand dune systems and inundation of salt marshes containing archaeological remains	Combination of sea-level rise and more frequent and severe storms	Negative Impact: Sites are uncovered and susceptible to being washed away						
Increased rainwater run-off and flooding	Increased erosion and/or deposition	Riverine waters carrying debris and sediment; changes to channels	Negative: site becomes susceptible to scour	Ecosystems changes from salt to freshwater	Flood waters remain	Negative Impact: Potential for sewage to increase microbiological activity due to nutrient enrichment Positive Impact: Changes to aerobic and anaerobic conditions which slow down degradation	Changes to aerobic and anaerobic degradation mechanisms through contamination	Increased heavy metal and other pollutants in sediments	Negative Impact: site moves from preservation state to active degradation with new chemical inputs.

Table 1.3: Matrix exploring the impact of warming seas on the physical, chemical, and biological process of shipwreck site formation. Copyright: Deanna Groom.

Input	Sea warming 1-4°C by 2100								
	Physical Processes Impact			Biological Processes Impact			Chemical Processes Impact		
	Process	Cause	Impact	Process	Cause	Impact	Process	Cause	Impact
Wider variability at sub-decadal scale in sea-surface and sub-surface temperatures (cooler periods and warmer periods)	Ocean's ability to absorb vast amounts of heat in one place and time and release it back into atmosphere in another place and time primary source of climate predictability	Control oceans exercise over atmospheric behaviour will become more variable with more extreme events such as hurricanes.	Sediment entrained by storms events leading to increased periods of uncovering, and vulnerability to physical force of storms	Substantial reduction in surface nutrients.	Increase in oceanic stratification	Changes in wreck ecology impacting on preservation			
Increases in monthly mean marine air temperatures (e.g., rate of 0.08°C decade since 1904, and at a rate of 0.4°C decade over the last 3 decades)							Temperature changes result in shifting chemical equilibria in the ocean acidification process (e.g., pH & carbonate content decrease)	Seawater temperature regulates ocean acidification and redistribution of chemical species (dissolved CO2, bicarbonate and carbonate ions, and hydrogen protons)	Increases in temperature increases the rates of chemical corrosion
Sea-level rise by thermal expansion and melting ice sheets increased water levels, storm surge and coastal flooding	Soft coastlines retreat in response	Erosion of softer cliff deposits and sand dunes frontages	Degradation of sites which had become relatively stable due to erosion; archaeological evidence washed away						

Input	Sea warming 1-4°C by 2100								
	Physical Processes Impact			Biological Processes Impact			Chemical Processes Impact		
	Process	Cause	Impact	Process	Cause	Impact	Process	Cause	Impact
Species adapted to live within narrow temperature ranges will be impacted by increased sea temperatures				Reduced reproduction rates of key species; mobile species move to better suited sea areas	Increase of sea temperature at rates predicted by RCP 8.5	Changes to wreck ecology providing opportunities for aggressively degrading species			
Impact on food supply chains	Commercial fish stock species move to new regions		Potential for sites to become impacted by fishing activities such as trawling and dredging	Impact on food supply for calcifying organisms	Die-off of cold-water corals (e.g., *Desmophyllum pertusum*) and reduced reef biodiversity	Reduction of protective biofouling layers for upstanding wood/iron/steel wrecks			
Increased de-oxygenation/ hypoxia events				Can cause mortality, reduced growth rates, and altered distributions and behaviours of species in affected waters and beyond	Eutrophication from nutrients (primarily N and P) and biomass from human waste and agriculture, as well as N deposition from fossil fuel combustion, stimulate the growth of algae	Changes to wreck ecology and protective biofouling layers; slow down degradation caused by fungi and bacteria in water column	Oxygen consumption at higher levels of the water column reduces dissolved oxygen at seabed level	Eutrophication from nutrients (primarily N and P) and biomass from human waste and agriculture, as well as N deposition from fossil fuel combustion, stimulate the growth of algae	Potential benefit: reduced amounts of dissolved oxygen available for degradation processes

Table 1.4: Matrix exploring the impact of acidification on the physical, chemical, and biological process of shipwreck site formation. Copyright: Deanna Groom.

Input	Increased Acidification								
	Physical Processes Impact			Biological Processes Impact			Chemical Processes Impact		
	Process	Cause	Impact	Process	Cause	Impact	Process	Cause	Impact
Overall trend for CO_2 uptake and magnitude of pH change moving away from base towards increasing acidity							Magnitude change in pH will increase mil/year rate of corrosion of metal wrecks	Uptake of CO_2 will continue to increase unless worldwide greenhouse gas emissions are brought under limits advised by IPCC	Negative Impact: Metal wrecks will be structurally weakened at a slightly faster rate leading to collapse and loss of evidence
Increase in seasonal variability in pH in open waters of Continental shelf							Magnitude change in pH will increase mil/year rate of corrosion for metal wrecks, particularly in the Spring and Summer months	Uptake of CO_2 will continue to increase unless worldwide greenhouse gas emissions are brought under limits advised by IPCC	Negative Impact: Metal wrecks will be structurally weakened at a slightly faster rate leading to collapse and loss of evidence
In near-coastal and estuaries of lower and intermediate salinity, CO_2 uptake and magnitude of pH change is less well understood.							Potential magnitude change in pH will increase mil/year rate of corrosion of metal wrecks	Uptake of CO_2 will continue to increase unless worldwide greenhouse gas emissions are brought under limits advised by IPCC	Negative Impact: Metal wrecks will be structurally weakened at a slightly faster rate leading to collapse and loss of evidence
Impact on the shell and skeleton building species within wreck's ecosystem from Calcium carbonate ($CaCO_3$) minerals, such as aragonite and calcite, being more prone to dissolution				Decline in biogenic calcification and calcium carbonate preservation, along with extinction events for marine calcifiers	Species using Calcium carbonate ($CaCO_3$) minerals in shells and skeletons will need to use more energy in constructing shells and skeletons	Negative Impact: Biofouling, which has some benefit for protecting surfaces from mechanical abrasion, will be reduced, with adverse consequences for the wreck's ecosystem causing a change in preservation state			

The matrices presented above reveal the complexity of the ocean and climate science involved in understanding climate change impacts on shipwreck sites. The matrices have demonstrated gaps where more research is needed, particularly regarding identifying thresholds for potentially rapid change in preservation states in relation to future climate change predictions.

Results of GIS Mapping

A 4km-by-4km sampling grid was generated in ArcGIS to cover the study area, involving the spatial area defined in the Welsh National Marine Plan. The attributes of each grid square collate information about present-day environmental parameters (e.g., salinity, temperature), predictions for change to these by the year 2050, and with a value assigned to represent the likely preservation qualities of the sedimentary regime. Wave data was acquired from the four semi-permanent WaveNet buoys to explore minimum, maximum and average depths for the Storm Wave Base. The four datasets span seventeen years, from 2005-2022. A fifth percentile of the largest significant wave heights was used to calculate average minimum and maximum values over the sequences provided.

The topography of the Welsh coast creates very distinct wave climates and differences in the Storm Wave Base. For example, south Wales is fully exposed to the longest fetch of the prevailing winds (southwest) and the swell waves of the Atlantic. This is in contrast to the largest waves and surges which occur in Liverpool Bay due to westerly and north-westerly winds, whose longest fetch is contained within the Irish Sea. Liverpool Bay is sheltered from swell waves from the Atlantic and experiences locally wind-generated seas (Brown *et al.* 2010: 119). One depth does not easily fit the entire Welsh coast as, of course, it varies in each of the prevailing wave climates. The Storm Wave Base varied from 27m in Liverpool Bay to the north, 33m off northwest Wales and Anglesey, 42m in Cardigan Bay, 50m off south Pembrokeshire coast, and 33m in the Bristol Channel to the south (Figure 1.5). However, 370 confirmed wrecks were subsequently identified as frequently in a state of sediment motion during storm conditions. In relation to the good preservation conditions identified in the grid shapefile described previously, some 84 sites fall within grid squares which suggest there would otherwise by good preservation conditions (Figure 1.6).

Figure 1.5: Area of seabed frequently set in motion during bad weather and storm events (based analysis of 5th percentile of highest significant wave heights). Copyright: Deanna Groom.

Figure 1.6: Grid squares assigned with High (green) or Very High (yellow) Archaeological Potential based on environmental factors such as sediment type and lower current/tidal flows. The number of located wrecks each grid square contains are also given. Copyright: Deanna Groom.

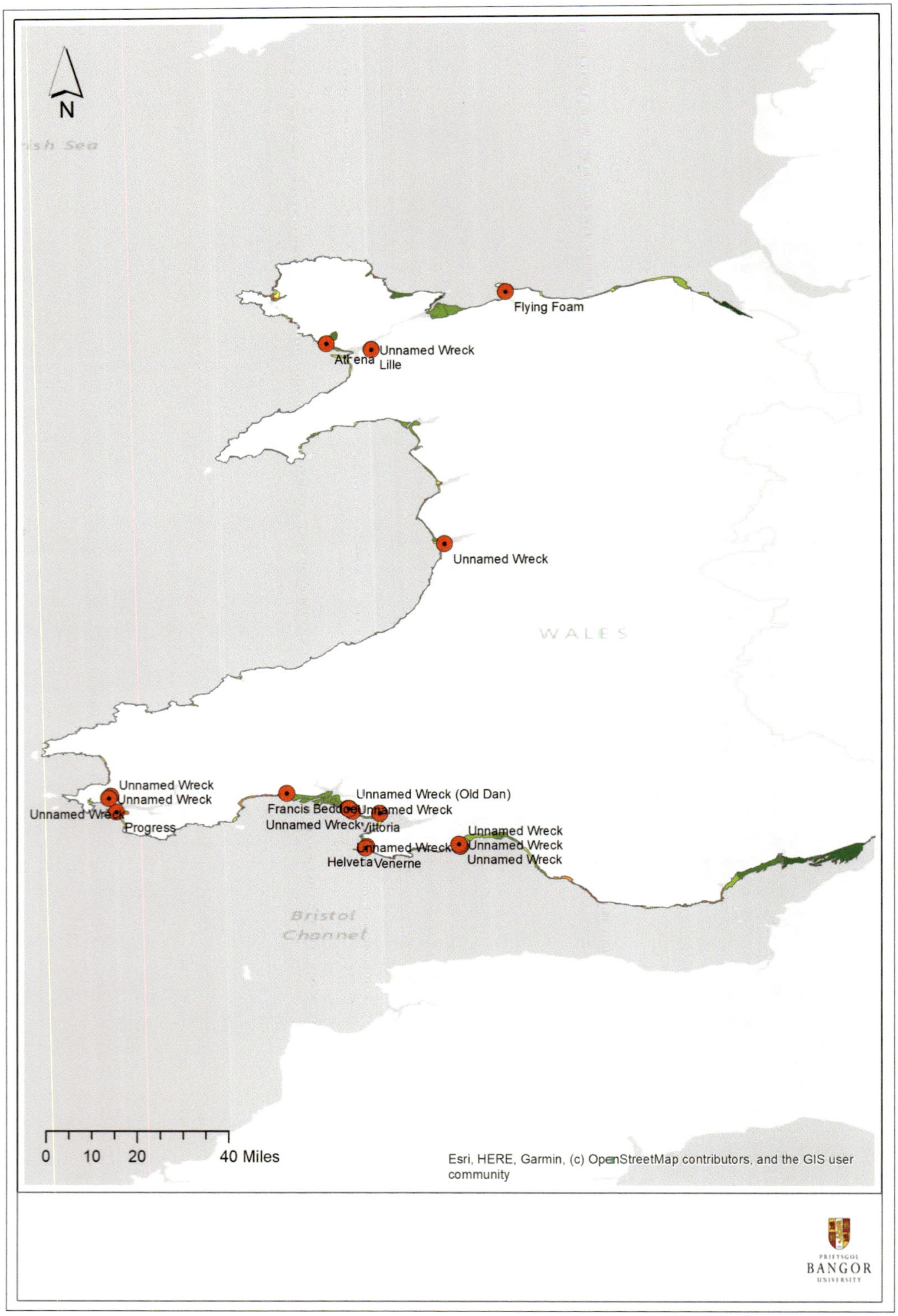

Figure 1.7: The location of intertidal wreck sites where photogrammetric studies were conducted. Coastal vulnerability analysis was created as a series of polygons with the best preservation conditions likely to find in stretches of coastline coloured green. Copyright: Deanna Groom.

To identify beaches where significant draw-down, in response to storminess, is a regular occurrence, the following geomorphological and marine environment characteristics were mapped to produce another shapefile:

- the character of the coastline and foreshore;
- the source of sediment (e.g., nature of cliffs or whether the sediment was a reworking of glacial tills or surface deposition in the nearshore);
- annual mean significant wave heights;
- offshore fetch;
- predictions for sea-level rise;
- net sediment transport;
- beach conditions;
- availability of beach profiles to provide evidence of trends.

This led to the effective identification of 77 beaches with a history of seasonal drawn-down. The wrecks within these beaches (Figure 1.7; Table 1.1) were prioritised for the monitoring programme of the project.

Results of Photogrammetric Field Survey

The monitoring programme was designed to first capture intertidal wrecks in their 'winter' profile. That is, after a prolonged period of successive severe weather events (i.e., visited in February-April and/or after a single particularly severe weather event). These surveys were then followed up by surveys capturing sites in their 'summer' profile (i.e., visited in August-early October after a prolonged period of calmer weather) to provide comparative data.

Photogrammetry was selected as the best methodological approach for these surveys, given the consistency of the methodology in depicting and reconstructing heritage assets, and also its efficiency in gathering data during short tidal windows (e.g., Lesgidi 2020; McCarthy 2014; McCarthy *et al.* 2019; Whitehead 2019). The images for the photogrammetry were taken with four low-cost waterproof cameras tested for the project, and Agisoft software was chosen for the 3D modelling of the sites (Figure 1.8). The photogrammetry was carried out through a process of establishing a perimeter around the centre of the site and capturing an image every five paces, at approximately 10m from the wreck or further away to capture the full extent of scour patterns. Then an inner circle of images was taken at one to two pace intervals and at a distance of 1m-1.5m from the wreck. Images of both the interior and exterior of the wrecks were captured, while a minimum overlap of 70% was observed to ensure the creation of successful photogrammetric 3D models. The number of images gathered for each site ranged between 160-760 (Figure 1.9). A Holystone HS720E 4K EIS drone (still resolution 4K - 3840 x 2160) was also used to facilitate access to locations where the foreshore conditions were uncertain. The drone was flown in straight line grid pattern to gather video and stills at a height of approximately 10m to facilitate total coverage of the intertidal sites. These aerial views proved particularly useful in the case of two new sites identified on the western bank of Sandy Haven. These were not immediately visible from the ground, as they were covered by seaweed, but were confirmed from the views gathered by the drone.

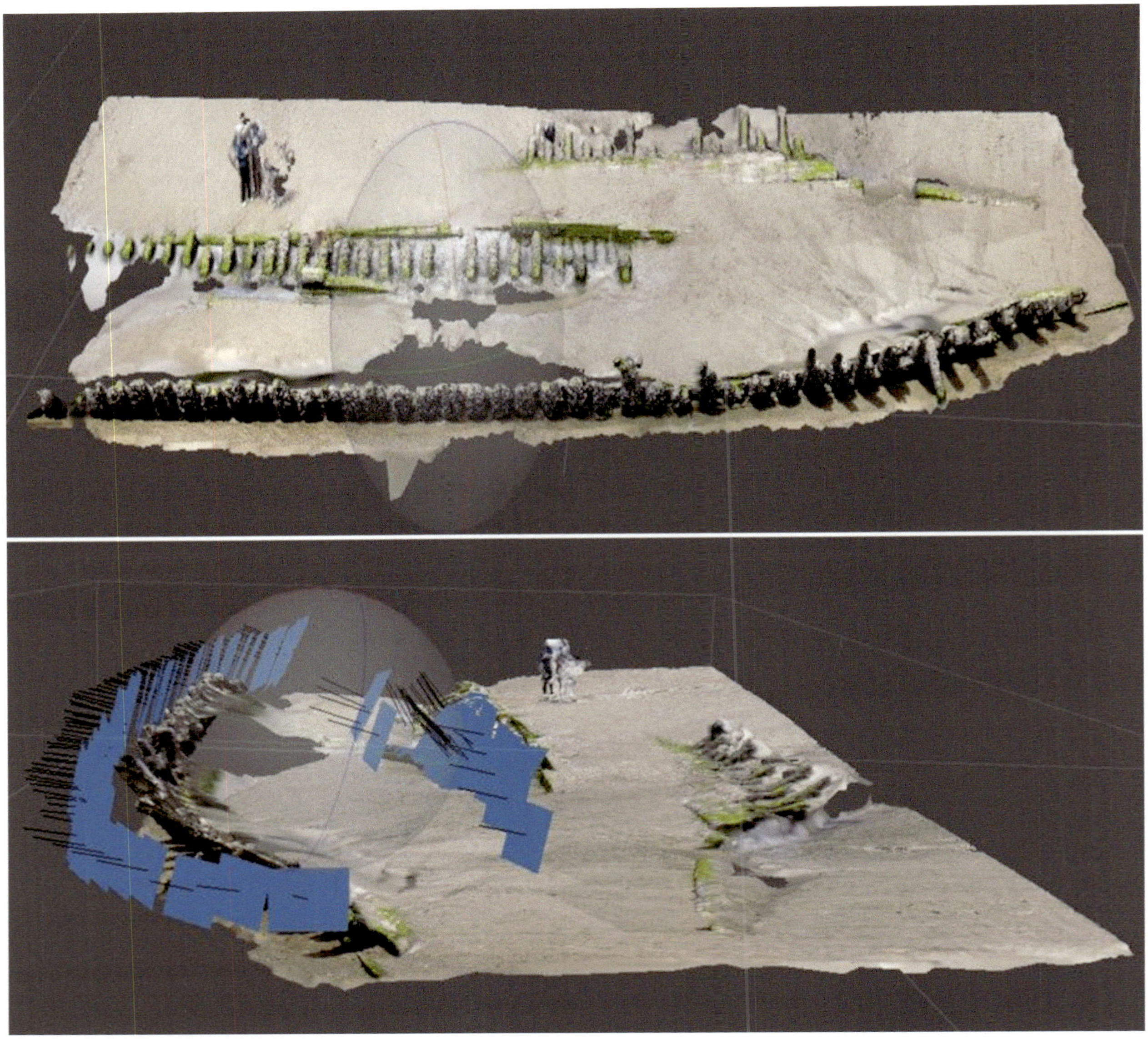

Figure 1.8: Above – the survey of the site of the Vittoria at Cefn Sidan, Carmarthen Bay, in June 2022, when the wreck becomes a popular landmark for beach walkers. Below – the 'Show Cameras' function of Agisoft reveals the pattern of photographs taken. Copyright: Deanna Groom.

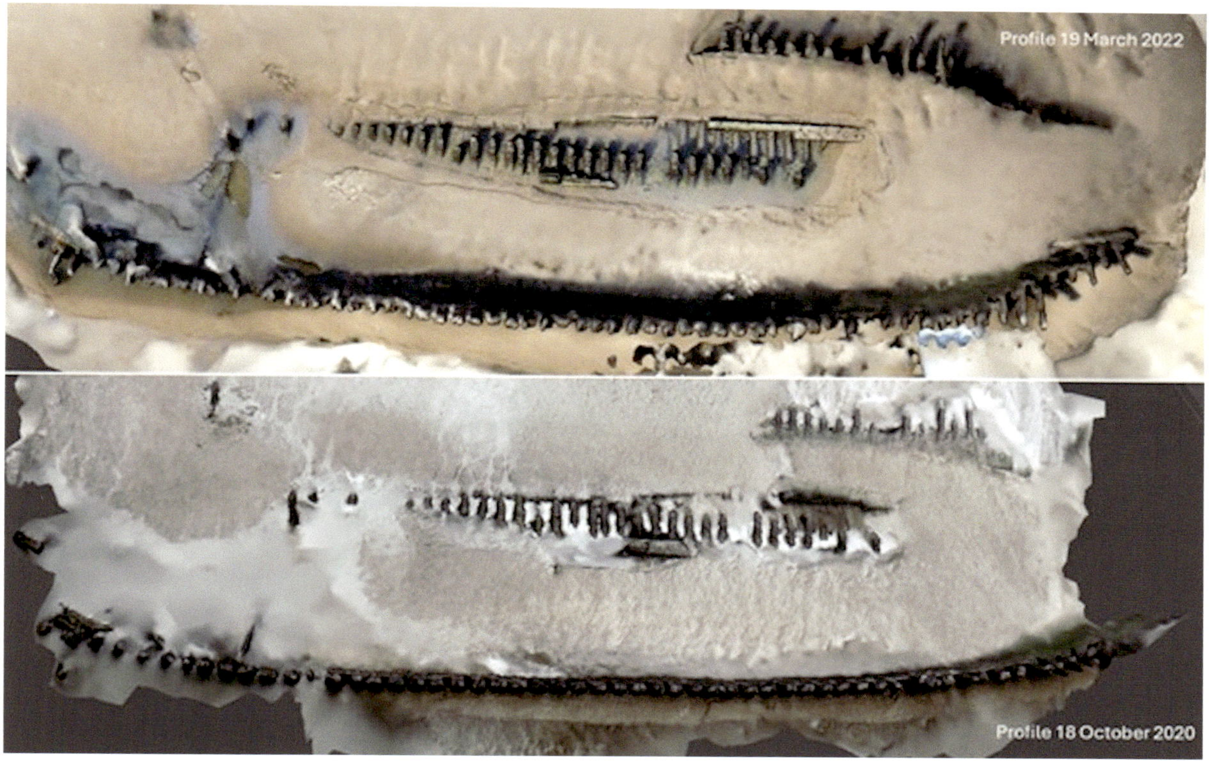

Figure 1.9: Photogrammetry surveys of the site, now believe to be the Vittoria, an Italian sailing barque which ran aground on 13 January 1872. The upper model was generated by images captured by the Olympus Tg-6 camera. The model below from images taken by a GoPro9. Copyright: Deanna Groom.

During the 3D modelling process, alignment issues occurred in the Agisoft software, where the surfaces of the scour pools were affected by wind ripples. To deal with this issue, a system of 'benchmark views' for each wreck was established, to include views of a folding 1m rules measuring the depths of scour pools. In other surveys, where the scour pool surfaces were still, Agisoft was able to provide an approximation of the depth and shape of the scour. Through a combination of these methods, it was found that scour depths varied seasonally across the 17 sites monitored for the project, from 0mm (infilled) in summer to 480mm deep after the winter storm period.

Cloud Compare software was used to assess the height changes of sediment cover across whole sites (Figure 1.10). These 3D models were supplemented by beach profile data provided by the Welsh Coastal Monitoring Centre gathered for coastal engineering purposes. Sites were located along the profile to establish the zone in which they fell (e.g., stable zone or primary area of beach drawn down/rebuild). The results of this analysis of the beach profiles at Pembrey are shown in Figure 1.11, with the relative locations of shipwreck sites marked along their lengths.

An analysis of modelling kindly supplied by the British Oceanographic Centre, Liverpool, reveals subtle differences observable in the Significant Wave Height distribution plots for the present (2005-2025 in blue) to the future (2050-2069 in yellow) (Figure 1.12). The plots suggest more Significant Wave Heights above 3.5m (i.e., more storm events in 2050-2069 period) and less calmer days with Significant Wave Heights in the range of 0.75m in North Wales and 1.5m in South Wales and Cardigan Bay (in the 2050-2069). The same data viewed as line plots suggest a quietening from June to mid-August, when the beach is most likely to rebuild with sediment brought back onshore. However, from north Wales around the coast to the south, in the wave distribution plots increases are evident in the wave climates up to 0–1m, suggesting fewer calm days. There are also increases in the Significant Wave Heights above 2.5m confirming predictions for more severe storm events.

The net result is that intertidal archaeological sites will be subject to increased erosion and physical force during autumn, winter, and spring. There will be a shorter, calm period in the summer for sediment to move back onshore. This means the sites will be uncovered and vulnerable to degradation for longer.

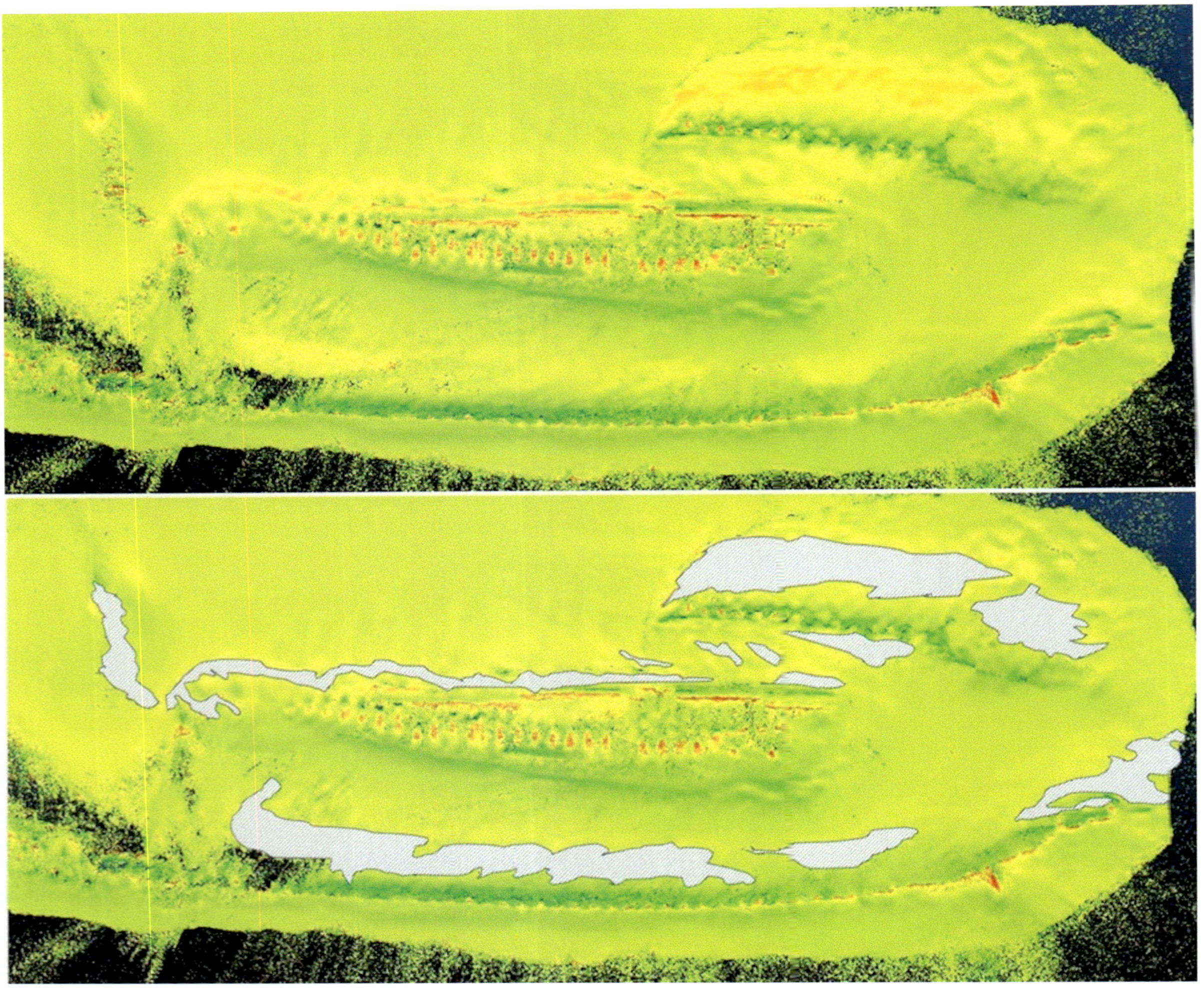

Figure 1.10: Agisoft point clouds from the two surveys were brought into Cloud Compare to reveal areas where erosion (green) and accretion (yellow-orange) had occurred. Copyright: Deanna Groom.

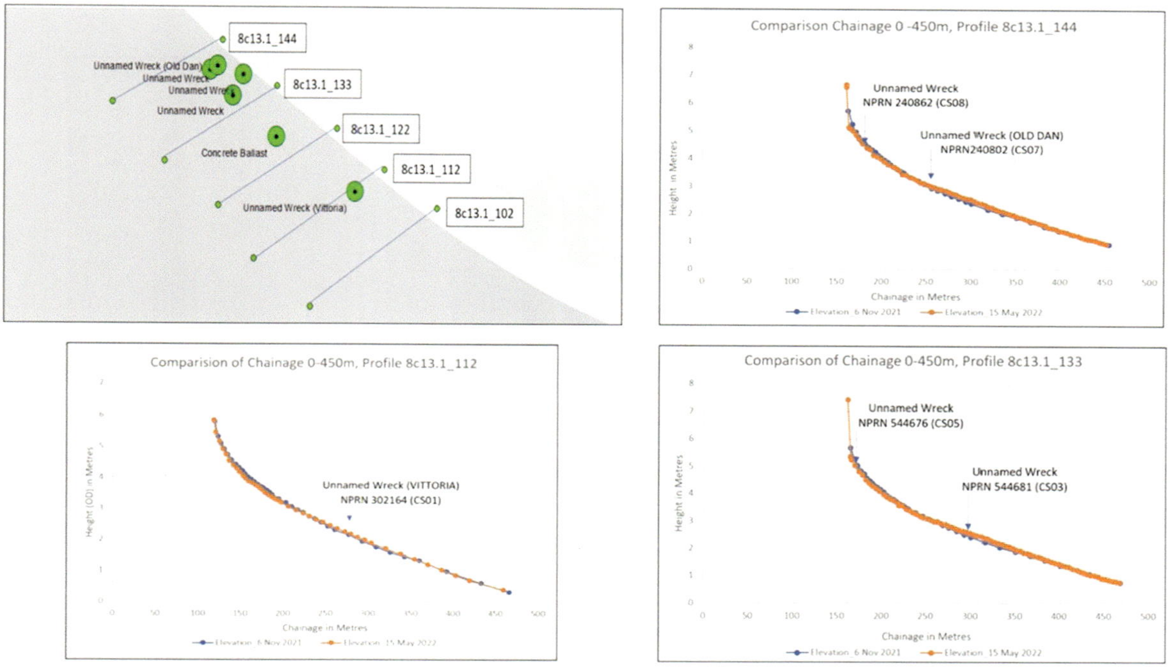

Figure 1.11: The beach profiles in the closest vicinity to the wrecks in the foreshore of Cefn Sidan reveals that the sites are within an area of the foreshore that draws down and rebuilds. Copyright: Deanna Groom.

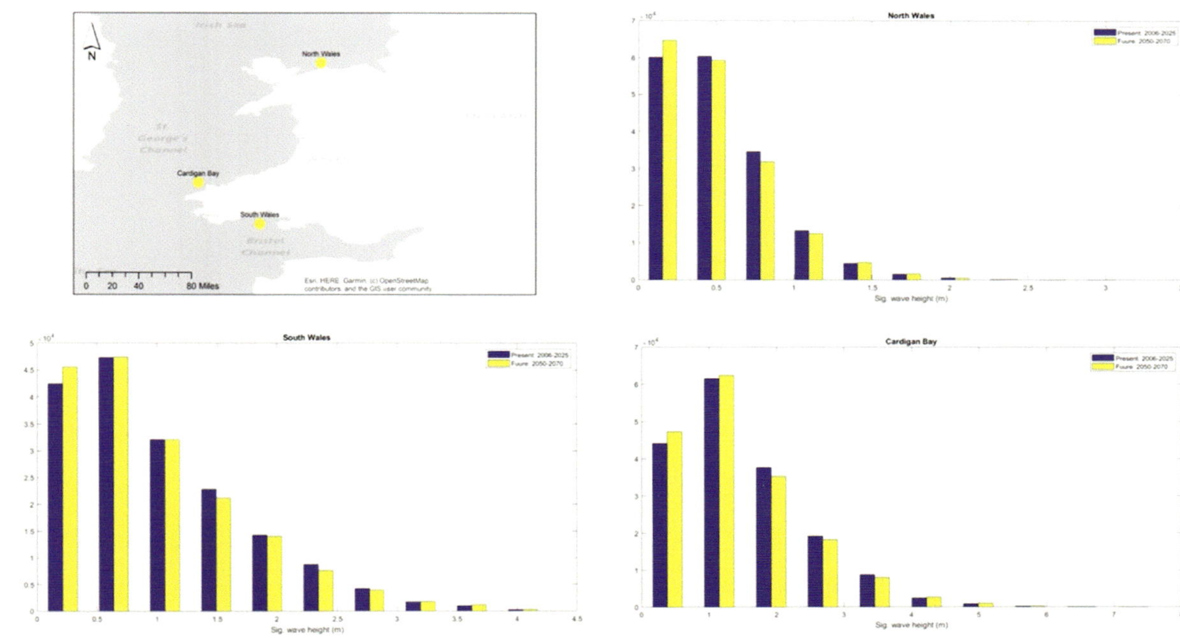

Figure 1.12: Distribution plots of present (2006-2025) and future (2050-2070) Significant Wave Heights kindly supplied by National Oceanographic Centre, Liverpool. Copyright: Deanna Groom.

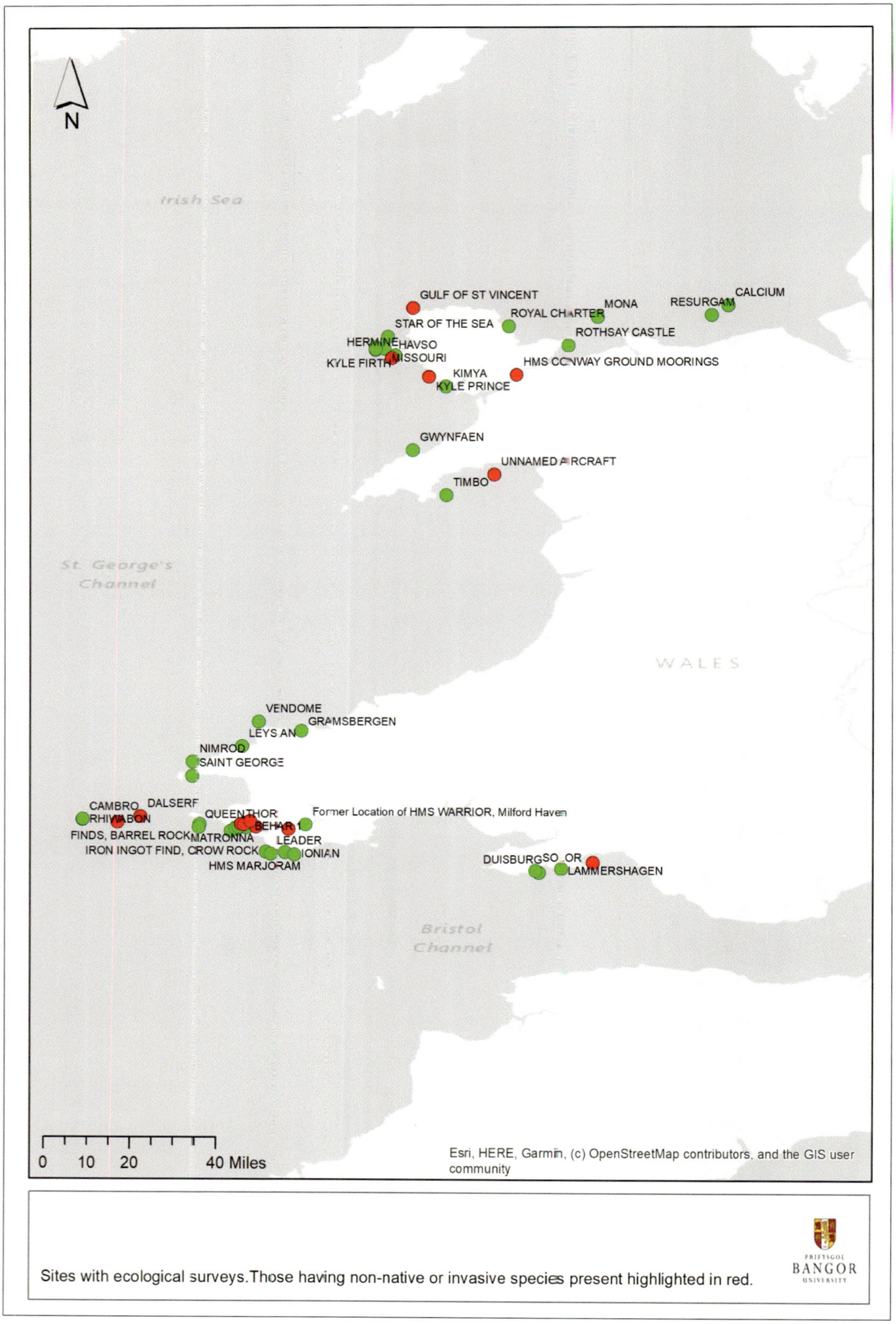

Figure 1.13: Wreck sites with ecological surveys in the data by the Marine Conservation Society including a Designated Historic Shipwreck - the early submarine Resurgam off Rhyl. Copyright: Deanna Groom.

Analysis of Ecological Surveys

From ecological surveys of 52 wreck sites, data for marine species and their abundance was compiled into a Microsoft Access database. The ecological surveys were carried out by members of the Marine Conservation Society as part of their 'Seasearch' initiative and also part of the data holdings of the National Biodiversity Network. For the present study the site species information was cross referenced to the appropriate monument record. 526 ecological records were created comprising 478 individual species and 48 biotopes (i.e., habitats associated with a particular community of species). These were enhanced with species information gleaned from a variety of published and online sources.

From this research, 14 shipwreck sites were found to feature one or more of seven non-native/ invasive species (Figure 1.13). The introduction of two of these new species, the *Ciona robusta* (yellow-ringed sea squirt) and the *Aplidium glabrum* (honeycomb sea squirt), is attributed to climate change and warming seas (Picton and Morrow 2023). A Second World War landing craft (Coflein, NPRN 240258) which foundered in 1951 within the port of Milford Haven has the highest number of identified invasive species (five in total).

In the UK, the Joint Nature Conservation Committee (JNCC) has promoted the use of a unified system for recording the abundance of marine benthic flora and fauna in ecological surveys, known as the SACFOR system (Hiscock 1996). It uses quadrants to undertake species counts and a descriptive scale which has given the system its name (**S**uper-abundant, **A**bundant, **C**ommon, **F**requent, **O**ccasional, and **R**are). The widespread adoption of SACFOR provides a means to compare both the presence and density of species from one survey to the next. Analysis of the SACFOR abundance attributed to each species at time of survey, identified 87 species with abundances in decline, whilst the populations of 80 others were found to be increasing. Sixty species were found to have stable abundance levels. Overall, analysis suggests that biodiversity on the 52 wreck sites studied in this project has declined by 7.9% since 1995. This corresponds with the trend identified in the State of Nature Report published in 2023, which suggests that Welsh wildlife numbers have decreased on average by 20% since 1994 (State of Nature – Wales 2023).

The Marine Life Information Network (MarLIN) provides methodologies for assessing species sensitivity to various physical, chemical, and biological pressures (Tyler-Walters *et al.* 2023). Using these, and identifying the factors most relevant to climate change, this part of the research identified 29 species with the potential biomarkers for climate change impact, from changes to the geographical range species are known to inhabit (Table 1.5). For example, the first UK sightings of the *Raja microocellata* (small-eyed ray) was in 2007. The ray was seen near the wreck of a small World War I submarine (HMSE39) at the mouth of Milford Haven. The numbers caught off the southwest coast of the UK and Ireland (during recreational fishing or by commercial fishing) has increased in the last two decades. According to research available on FishBase (Frose and Pauly 2024), it is believed that this species will continue to expand its range into the North Sea.

The study of ecological surveys in relation to underwater and intertidal archaeological remains in Wales have been extremely insightful for understanding changes in shipwreck ecology. However, these surveys pose some limitations. First, the vast majority of sites are nearshore in depths of less than 30m. There is no comparative data for sites in the deeper parts of Welsh waters (i.e., Celtic Deep, St George's Channel and the Irish Sea). Moreover, the species recorded are 'macro' flora and fauna. They do not include microbiological species, such as the marine fungi and bacteria, which are known to play a key part in shipwreck degradation. In particular, marine wood-boring species, such as gribble and shipworm, play a major role in weakening and breaking down wooden shipwrecks.

However, our knowledge of present range and where they might expand in numbers with warming seas is extremely poor. Even though it is known that the invasive gribble *Limnoria tripunctata*, which favours sea temperatures of 14°C and warmer, was first observed near Newport in 1951 and at Swansea in 1959 (Jones 1963), no recent sightings have been recorded in recent ecological surveys. There is only one other confirmed sighting for the UK for this gribble, from Portsmouth in October 2024 (National Biodiversity Atlas). But, as explained above, this absence might be due to lack of data rather than lack of northward gribble and shipworm expansion.

Table 1.5: Species with the potential to be used as climate change biomarkers selected from species present on the 52 underwater sites. Copyright: Deanna Groom.

Southern species currently whose extent of distribution or abundance might increase	Northern species which may either decrease in abundance and extent or disappear
Amathia citrina	*Urticina felina*
Aplidium glabrum	*Henricia sanguinolenta*
Axinella dissimilis	
Balistes carolinesis	
Callionymus reticulatus	
Crisualria plumosa	
Drachiella heterocarpa	
Entelurus aequoreus	
Gymnangium montagui	
Haliclona simulans	
Halopteris filicina	
Halyphysema tumanowiczii	
Homaxinella subdola	
Jassa falcata	
Kallymenia reinformis	
Macropodia deflexa	
Raja microcellata	
Palinurus elephas	
Parazoanthus axinellae	
Perophora listeri	
Phorbas fictitius	
Stolonica socialis	
Taonia atomaria	
Thecacera pennigera	
Zeus faber	

Conclusion

From this study it becomes clear that the life of a ship is not over when it sinks or runs aground. Shipwreck sites affect the structure, chemistry, and biological makeup of marine ecosystems and they are equally impacted by changes that take place in their surrounding environment. Climate change is certain to impact the longevity of many underwater and intertidal cultural heritage assets, putting many shipwrecks in Wales and worldwide at risk.

The practitioners' survey carried out as part of this project has confirmed that there are many areas of uncertainty which deserve future interdisciplinary research and worldwide collaboration. For example, the preferred *in-situ* preservation of underwater cultural heritage assets requires radical review in the face of the climate emergency, since various climate change parameters have already put submerged and intertidal archaeological sites at risk.

In Wales, monitoring programmes have already begun to shed light on the changes taking place with the changing climate. As part of these efforts, the photogrammetric surveys of intertidal wrecks in their winter and summer profiles, included in this project, have demonstrated significant potential for gathering environmental and heritage data quickly and efficiently, allowing thus long-term monitoring of sites. Moreover, the interdisciplinary GIS shapefiles created in this research, have provided quick access to a wide range of spatial and environmental information that could assist in the identification of best preservation condition for heritage assets that might be at risk. The ground-truthing of the analysis presents the potential to take the results forward in collaboration with marine geologists utilising a combination of geotechnical data (cores) and radicisotope analysis to the identify storm event horizons in sediment layers.

Moreover, the absence of data for the abundance of species of shipworm and gribble around the Welsh coast requires urgent further attention. The methodologies developed for exploring the impact of climate change on commercial fish stocks might be applied to the full range of species living on underwater archaeological sites with the assistance of marine biologists.

Overall, as this study highlights, the key to fully understand the impacts of climate change on shipwreck sites is the encouragement of international cross-disciplinary studies combined with stakeholder collaboration. The objective is to create the best circumstances to mitigate risks and safeguard our common underwater and intertidal cultural heritage assets for the future.

Acknowledgements

A special thank you is extended to Dr Trevor Winton, Flinders University, for assistance with the matrices and to the Royal Commission on the Ancient and Historical Monuments of Wales for allowing use of the shipwreck dataset from the National Monuments Record of Wales.

References

Bricheno, L. M. and J. Wolf. 2018. Future Wave Conditions of Europe, in Response to High-End Climate Change Scenarios. *Journal of Geophysical Research: Oceans* 123.12: 8762–8791. <https://doi.org/10.1029/2018JC013866>

Brown, J. M., A. J. Souza and J. Wolf. 2010. An investigation of recent decadal-scale storm events in the eastern Irish sea. *Journal of Geophysical Research: Oceans* 115.5: 1–12. <https://doi.org/10.1029/2009JC005662>

Coflein, online database for the National Monuments Record of Wales (NMRW), Royal Commission on the Ancient and Historical Monuments of Wales (RCAHMW), viewed 15 March 2025, <https //coflein.gov.uk/en/>.

Couple Model Intercomparison Project 5 (CMIP 5), viewed 15 March 2025, <https://wcrp-cmip.org/cmip5/>.

Dunkley, M. 2015. "Climate is what we expect, weather is what we get": managing the potential effects of oceanic climate change on underwater cultural heritage, in W. J. H. Willems and H. P. Van Schaik (eds) *Water and heritage: material, conceptual, and spiritual connections*: 217–230. Amsterdam: UNESCO.

Froese, R. and D. Pauly. 2024. FishBase, viewed 15 March 2025, <www.fishbase.org>.

Gearheart, R., D. Jones, A. Borgens, S. Laurence, T. DeMunda and J. Shipp. 2011. *Impacts of Recent Hurricane Activity on Historic Shipwrecks in the Gulf of Mexico Outer Continental Shelf.* New Orleans: U.S. Department of the Interior - Bureau of Ocean Energy Management, Regulation and Enforcement Gulf of Mexico OCS Region.

Gjelstrup Bjordal, C., D. Gregory and A. Trakadas. 2011. *Wreck Protect: Decay and protection of archaeological wooden shipwrecks.* Oxford: Archaeopress.

Hiscock, K. 1996. *Marine Nature Conservation Review: Rationale and methods. Coasts and seas of the United Kingdom.* Peterborough: Joint Nature Conservation Committee.

Howarth, M. J. 2005. *Hydrography of the Irish Sea: SEA6 Technical Report.* London: British Geological Survey and UK Department of Trade and Industry's offshore energy Strategic Environmental Assessment programme.

Humphreys, M. P., Y. Artioli, D. C. E. Bakker, S. E. Hartman, F. León, S. Wakelin, P. Walsham and P. Williamson 2020. Air – sea CO 2 exchange and ocean acidification in UK seas and adjacent waters. *Marine Climate Change Impacts Partnership Science Review*: 54–75. <https://doi.org/10.14465/2020.arc03.oac>

ICOMOS (International Council on Monuments and Sites) 1997. The ICOMOS Charter on the Protection and Management of Underwater Cultural Heritage. *Conservation and Management of Archaeological Sites* 2.1: 35–38. <https://doi.org/10.1179/135050397793138925>

Jarvis, C., M. P. Ermida and O. Varmer 2023. Threats to Underwater Cultural Heritage from Existing and Future Human Activities. *Blue Papers* 2.1: 76–83. <https://doi.org/10.58981/bluepapers.2023.1.08>

Jarvis, C. 2024. *Threats to Our Ocean Heritage: Bottom Trawling.* Cham: Springer. <https://doi.org/10.1007/978-3-031-57953-0>

Jones, L. T. 1963. The geographical and vertical distribution of British Limnoria [Crustacea: Isopoda]. *Journal of the Marine Biological Association of the United Kingdom* 43.3: 589–603. <https://doi.org/10.1017/S0025315400025546>

Kintisch, E. 2016. Arctic shipworm discovery alarms archaeologists. *Science* 359.6276: 901. doi: 0.1126/science.351.6276.901. <https://doi.org/10.1126/science.351.6276.901>

Lesgidi, P. 2020. Interpreting the Significance of Underwater Archaeological Photogrammetry. *Maritime Archaeology Graduate Symposium 2019.* London: Honor Frost Foundation. <https://doi.org/10.33583/mags2019.07>

Manders, M. 2008. *In Situ* Preservation: “the preferred option”. *Museum International* 60.4: 31–41.

Marine Life Information Network (MarLIN), viewed 15 March 2025, <https://www.marlin.ac.uk/>.

McCarthy, J. 2014. Multi-image photogrammetry as a practical tool for cultural heritage survey and community engagement. *Journal of Archaeological Science* 43.1: 175–185. <https://doi.org/10.1016/j.jas.2014.01.010>

McCarthy, J., J. Benjamin, T. Winton and W. Van Duivenvoorde 2019. The Rise of 3D in Maritime Archaeology, in J. McCarthy, J. Benjamin, T. Winton and W. Van Duivenvoorde (eds) *3D Recording and Interpretation for Maritime Archaeology.* Cham: Springer. <https://doi.org/10.1007/978-3-030-03635-5_1>

Murphy, P., C. Pater and M. Dunkley. 2008. Out to Sea: climate change and the maritime historic environment. *Conservation Bulletin* 57 Spring: 17–19.

Murphy, P., D. Thackray and E. Wilson. 2009. Coastal Heritage and Climate Change in England: Assessing Threats and Priorities. *Conservation and Management of Archaeological Sites* 11.1: 9–15. <https://doi.org/10.1179/135050309X12508566208281>

Mustow, S. E. 2021. Marine Invasive Non-native Species in the UK: Scale of the Problem and Progress of the Response. *CIEEM in practice* 112: 21–25.

National Biodiversity Atlas, viewed 15 March 2025, <https://nbnatlas.org/>.

National Biodiversity Network, viewed 15 March 2025, <https://nbn.org.uk/>.

Paxton, A. B., C. Mcgonigle, M. Damour, G. Holly, A. Caporaso, P. B. Campbell, K. S. Meyer-Kaiser, L. J. Hamdan, C. H. Mires, and C. Taylor. 2024. Shipwreck ecology: Understanding the function and processes from microbes to megafauna. *BioScience* 74.1: 12-24. <https://doi.org/10.1093/biosci/biad084>

Perez-Alvaro, E. 2016. Climate change and underwater cultural heritage: Impacts and challenges. *Journal of Cultural Heritage* 21: 842–848. <https://doi.org/10.1016/j.culher.2016.03.006>

Picton, B. E. and C. C. Morrow, 2023, Encyclopedia of Marine Life of Britain and Ireland – Habitas, viewed 15 March 2025, <https://www.habitas.org.uk/marinelife/>.

Seasearch, Marine Conservation Society, viewed 15 March 2025, <https://www.seasearch.org.uk/>.

Sharples, J., J. Holt and S. Wakelin 2020. Impacts of climate change on shelf-sea stratification, relevant to the coastal and marine environment around the UK. *Marine Climate Change Impacts Partnership Science Review*: 103-115. <http://dx.doi.org/10.14465/2020.arc05.str>

State of Nature – Wales, 2023, viewed 15 March 2025, <https://stateofnature.org.uk/wp-content/uploads/2023/09/TP26053-SoN-Wales-summary-report-v10.pdf.pagespeed.ce.UcI3aoHAY6.pdf>.

Stieglitz, T. C. and P. Waterson 2013. Impact of Cyclone Yasi on the wreck of the SS Yongala documented by comparative multibeam bathymetry analysis *Queensland Archaeological Research* 16: 33–43. <https://doi.org/10.25120/qar.16.2013.222>

Tinker, J. and E. L. Howes. 2020. The impacts of climate change on temperature (air and sea), relevant to the coastal and marine environment around the UK. *Marine Climate Change Impacts Partnership Science Review*: 1-30. <https://doi.org/10.14465/2020.arc01.tem>

Tyler-Walters, H., H. M. Tillin, E. A.S. d'Avack, F. Perry and T. Stamp 2023. Marine Evidence–based Sensitivity Assessment (MarESA) - Guidance Manual. *Marine Life Information Network (MarLIN)*. Plymouth: Marine Biological Association of the UK. <https://www.marlin.ac.uk/publications>

UNESCO, 2001, Convention on the Protection of Underwater Cultural Heritage, viewed 14 April 2024, <https://unesdoc.unesco.org/ark:/48223/pf0000126065>.

Wavenet, Cefas, viewed 15 March 2025, <https://wavenet.cefas.co.uk/map>.

Welsh Coastal Monitoring Centre, viewed 15 March 2025, <https://www.wcmc.wales/>.

Whitehead, H. W. 2019. Lost at Sea: Utilizing Photogrammetry to monitor two submerged F8F Bearcats. University of West Florida. Unpublished MA Dissertation, University of West Florida.

Young, E. F. and J. T. Holt 2007. Prediction and analysis of long-term variability of temperature and salinity in the Irish Sea. *Journal of Geophysical Research: Oceans* 112.1. <https://doi.org/10.1029/2005JC003386>

2.
Climate Change and Intertidal Archaeology in the Basque Country (Spain): Looking at Possible Research Directions

José Manuel Matés Luque

Climate change effects such as sea level rise and extreme storminess are already affecting intertidal archaeological sites around the world, including the Basque Country (Spain), where much of intertidal archaeology has not been researched and recorded. By the year 2100, sea level rise of 140cm to 154cm has been predicted in recent models for some parts of the region (Sainz de Murieta *et al.* 2018). With these predictions beaches will disappear or be reduced in width to half their size, while the speed of currents in estuaries will increase affecting both the canal system and the riverbanks (Chust *et al.* 2013: 35). The remains of wrecks in the intertidal and estuary zones, currently at risk, offer a unique opportunity to record and understand maritime material culture of the region, before they are damaged by climate change effects or are permanently submerged. At the same time, wooden wrecks offer ideal sites to monitor and collect local climate and environmental data in these vulnerable and dynamic maritime zones. This paper explores the various research possibilities for intertidal archaeology in the Basque Country to protect heritage at risk while gathering invaluable localised scientific data that can assist in climate action, sustainability and climate resilience.

Introduction

Climate change is already affecting every part of the world. Droughts, floods, wildfires, extreme weather patterns including increased storminess, abnormally strong winds, waves, and tides, are impacting all organisms and ecosystems on Earth, which in return affects human life. In the 2010s most countries around the world set up national and international policies to tackle climate change and its negative impacts (e.g., The Paris Agreement 2015) with the goal to achieve sustainable development (see UN Sustainable Development Goals). In response to that, studies on threats to maritime heritage due to climate change effects (e.g., Baika 2012; Dunkley 2015; Gregory *et al.* 2022; Hafner and Underwood 2020; Kolaiti and Mourtzas 2020; Peev *et al.* 2020; Pérez-Álvaro 2016) including increase in sea temperature, sea level, pH level on metal wrecks, salinity, acidification, and shipworms due to warmer waters, known already to biologists, oceanographers, coastal scientists, geographers and other physical science researchers (e.g., Kersting 2016: 14-41), attempt to improve management of underwater, coastal and intertidal heritage assets.

The Basque country (Figure 2.1), on the north coast of Spain, has suffered from climate change effects, like most countries with coastal zones. Flooding in particular (Ibisate González de Matauco *et al.* 2000) has been very impactful and catastrophic, especially when heavy rains combined with high tides increase the volume of water at the riverbanks, which then suffer from extreme inundation, frequently resulting in loss of human lives and destruction of local livelihoods (e.g., Martínez Sarracina 2023). The Basque central government, along with other local and regional authorities, are aware of the impacts of climate change on the region and its citizens and have taken measures for climate resilience and adaptation (Gobierno Vasco 2015; Ihobe 2019).

Figure 2.1: Map of the Basque country, on the north coast of Spain. Copyright: Geovisor Euskadi, modified by J M Matés Luque.

This paper explores possible directions of intertidal archaeology and heritage initiatives in the Basque country that could contribute to the protection of maritime heritage vulnerable to climate change, while also assisting in sustainability and climate resilience efforts in the region. After briefly introducing past entanglements of natural and cultural heritage in intertidal projects of the Basque country along with the current climate change policies of the region, the paper proceeds with a presentation of the main threats that climate change poses to intertidal wrecks of the Basque country. Through selected examples, this analysis recommends routes of research that could be undertaken to promote integrated cultural and natural heritage research in the region to monitor heritage at risk while also collecting significant place-based climate and environmental data.

Environment and heritage in the Basque Country: past and present actions

In the past 30 years there have been efforts to combine natural and cultural heritage approaches in the intertidal zones of the Basque Country, even though not directly aimed at addressing climate change and environmental deterioration. For example, some of the earliest intertidal archaeological work in the Basque country took place in 1998 with a watching brief during works on the natural old canal of the river Oka, in Gernika, as part of a restoration initiative to revive local flora and fauna in the area. This project, even though nature based, uncovered various archaeological artefacts as well as parts of a 15th-century clinker-built boat (Pujana and Díez 1999). The significance of the discovery, a boat abandoned and buried in the old riverbank close to the trading town of Gernika, encouraged local stakeholders to set up a rescue team to excavate, record, and lift the remains of that boat (Izaguirre and Valdés 1998; Izaguirre *et al.* 1999, 2001). Similarly, research on various intertidal mills of the Basque region (Mansilla Hortigüela 2010; Tellería Sarriegi 2011, 2012) has brought together in insightful ways aspects of the local natural and historic environments and their significance in the dynamic space of the intertidal zone.

Currently, the Basque government recognises the importance of agroforestry and livestock industries as a tool to preserve cultural heritage (Gobierno Vasco 2015: 86). This concept, however, is related to traditional ways of living rather than any challenges with or threats to the preservation of archaeological remains. In fact, local policies and relevant scholarship related to climate change offers no mention to any cultural heritage (Gobierno Vasco 2015; Ihobe 2019) other than some brief notes on the need to reduce the environmental impact of museums, archives, orchestras, festivals and similar institutions (Gobierno Vasco 2022: 19), mostly focusing on the carbon footprint of arts and culture institutions.

Based on the existing frameworks and policies of the Basque Country, archaeology as a field, and maritime archaeology as a subdiscipline, have had no role in climate action, which is similar to policies of other countries around the world. Heritage is only mentioned in association with tourism and as part of raising awareness regarding sustainability, the protection of the environment and the need to reduce the carbon footprint particularly of travel (Gobierno Vasco 2022: 12). Similarly, reports by the Spanish Environmental Agency mention tourism as an element which will be affected by climate change but not explicitly heritage (Campos Rodrigues *et al.* 2017; Gómez Royuela 2016; Losada Rodríguez *et al.* 2014). Given that tourism is very important for the Spanish GDP, both in terms of revenue, and because of the summer jobs that it supplies (thus affecting employment), it is clear that climate change is a very important issue to consider. Similarly, reports on climate adaptation, published by local governments and related authorities in Spain focus on health, transport, and recommendations on how citizens could reduce the amounts of the water and energy they use, buy local products, reduce plastic packaging, reuse products and recycle as much as possible. All these efforts are very significant, but the scope of climate adaptation and pursuit of sustainable development can be far wider.

Ideas from maritime archaeological initiatives from across the world that address climate change and sustainable development could be applied in the Basque Country, thus providing local opportunities for more integrated protection of natural and cultural heritage, as recommended by the UN SDG Target 11.4 (UN SDGs), in the intertidal zone. Observations from fieldwork efforts of the last decade (e.g., Alonso Rodríguez and Matés Luque 2020; Matés Luque 2013, 2014, 2015, 2016a, 2016b, 2018a, 2019a, 2019b, 2020a, 2020b, 2020c) have demonstrated the significant potential of intertidal heritage of the Basque Country to provide local climate data but also deal with various obstacles that need to be overcome to ensure the protection of heritage that currently is or will come under threat.

Figure 2.2: Map indicating the wrecks recorded in the estuary of Lea (Lekeitio and Mendexa). The yellow arrow points to such wrecks. The red flag points out to the wreck in the middle of the canal, which was not authorised for further research and recording. Copyright: Geovisor Euskadi, modified by J M Matés Luque.

Figure 2.3: Map indicating the wrecks recorded in the estuary of Artibai (Ondarroa). Copyright: Geovisor Euskadi, modified by J M Matés Luque.

Looking at possible research directions for intertidal archaeology in the Basque Country

Until now, no project in the intertidal zone of the Basque Country has looked explicitly at climate change impacts on archaeological sites and particularly on boat remains. Hence, there is no relevant dataset or model that can assist in assessing risks to the historic environment and considering measures for safeguarding the respective heritage assets. The present study has used information available in the Basque Country's *Geovisor GeoEuskadi*. In that platform, anyone can activate and visualise in a georeferenced map layer different characteristics (e.g., place names, geology, land use, energy, transport, climate change effects such as flooding, archaeological sites, historical buildings, etc.). For the two main estuaries where wooden wrecks have been recorded by the author, the Lea estuary in the villages of Lekeitio and Mendexa (Figure 2.2), and the Artibai estuary in the village of Ondarroa (Figure 2.3), the layers regarding climate change effects, have been activated (Matés Luque 2019b, 2020a, 2022b).[1] By combining the climate change data available on *Geovisor GeoEuskadi* with the archaeological sites and remains recorded by the author, and guided by existing international scholarship on coupled climate-archaeology initiatives, it has been possible to assess potential threats to wooden wrecks of the intertidal zone at the selected region. At the same time, this study has explored potential opportunities for archaeology practitioners to offer data related to climate change in joint efforts with researchers from different fields, such as geomorphologists, biologists, environmentalists, oceanographers, coastal managers and similar. As it will be explained below, due to the elevated risk to heritage assets in the intertidal zone of the Basque Country, these interdisciplinary collaborations will offer the chance to mitigate risks, safeguard remains or provide preservation by record to sites that cannot be retained. Furthermore, the archaeological sites themselves can operate as place-based research grounds to monitor climate data, providing benchmarks and baselines to document how various elements of climate change are evolving at a local scale.

Threats to consider for place-based heritage safeguarding and climate change analysis

There are several threats to archaeological sites and heritage assets in the intertidal zones of the Basque Country. In what follows, the focus is primarily on those impacting wrecks and boat remains.

Sea-level rise

The melting of ice caps in the poles are increasing the volume of water in the oceans and causing sea-level rise around the world. In the next 100 years coastal and intertidal zones will be particularly affected, and according to various predictive models (e.g. Climate Central, Coastal Risk Screening Tool; NASA Sea Level Projection Tool) various regions will become submerged. For archaeological sites, this will mean that land-based sites will come closer to the water, previously intertidal material remains might become more frequently or permanently submerged, and the depth of already submerged archaeological deposits will increase (Wright 2016).

The Basque coast is known for its dramatic changes due to the impacts of the Atlantic Ocean. For example, its famous flysch rock formations have been formed over thousands of years by geological and hydrological pressures of the ocean, creating what it is called locally *rasa mareal* ('wave-cut platform' or 'abrasion platform' in English). The dynamic changes of the coast in the area, have also rendered some famous prehistoric caves and archaeological sites coastal, even though in prehistory, they would have been hundreds of meters away from the water. A Bizkaian example of this is the Santa Catalina cave in Lekeitio, currently coastal but in prehistoric times an inland site (Berganza Gochi and Arribas Pastor 2014). This dynamic change of the coast is related to the melt down of the ice cap about 12,000 years ago when the current coastline was shaped.

[1] Wrecks have been identified in other estuaries as well (e.g., Butrón estuary and the Nervión-Ibaizabal estuary), but, since the project is ongoing, they have not been included in this paper because the research has not been completed yet.

Figure 2.4: Area of the Lea estuary being affected by sea level rise. Yellow arrows indicate wreck sites. Copyright: Geovisor Euskadi, modified by J M Matés Luque.

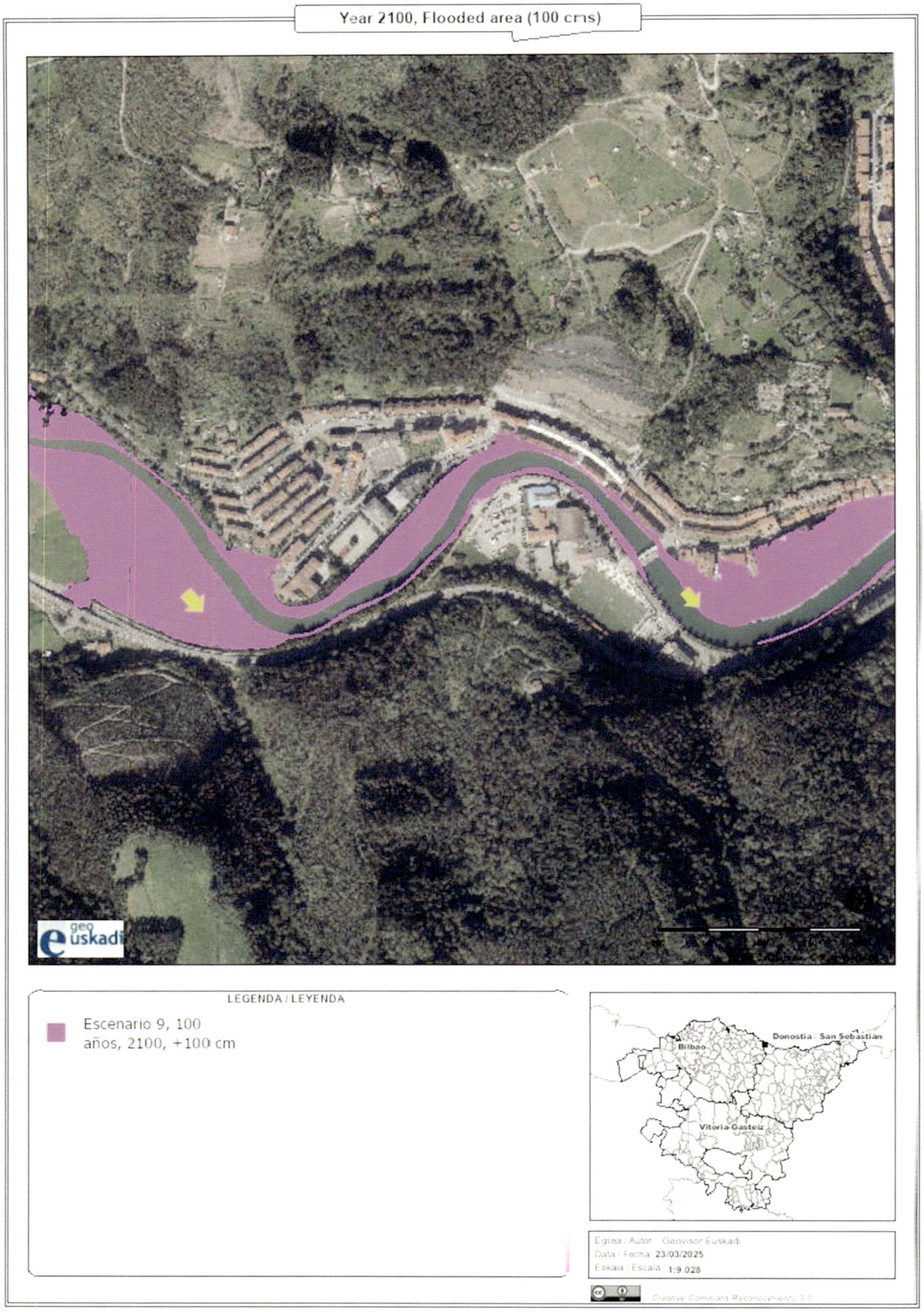

Figure 2.5: Area of the Artibai estuary being affected by sea level rise. Yellow arrows indicate wreck sites. Copyright: Geovisor Euskadi, modified by J M Matés Luque.

The future changes imposed by sea-level rise due to climate change are expected to affect the amount of archaeological data preserved as well as the conditions for safeguarding archaeological sites of the intertidal zones. By looking at the areas of the Lea and Artibai estuaries (Figures 2.4-2.5) that will be affected by sea-level rise, according to data available in the *Geovisor GeoEuskadi*, it becomes obvious that there are several shipwreck sites that will be impacted. Therefore, currently there is a limited time to record heritage remains, such as the wooden barge in Lekeitio shown in Figure 2.6, or the watercraft at Ondarroa shown in Figures 2.7 and 2.8, that might be damaged or lost forever. Introducing a climate change angle to archaeological watching briefs, undertaken as part of development projects and civil engineering works, similar to work done already by the author (e.g., Matés Luque 2018b, 2022a), would be an easy way to enhance our archaeological knowledge of these at-risk areas. Preservation by record, especially with frequent photographic or photogrammetric surveys could enable the preservation of threatened sites and also the real-time monitoring of heritage loss. At the same time, monitoring sea-level rise on heritage sites can offer an opportunity for non-heritage professionals to collect local climate data for the protection of other, non-archaeological human material culture, and particularly contemporary human settlements, that might come under threat in the following decades.

Figure 2.6: Wooden barge in Lekeitio (Lekeitio I). The boat is affected by the tide and silt of the river. Currently, the stempost is covered with soil and colonised by plants. Copyright: J M Matés Luque.

Figure 2.7: Ondarroa I fishing ship. This vessel is covered by a thick and hard layer of debris. Due to its location very close to the mouth of the estuary, recording is difficult with a very limited tide window. Eventually, this wreck will be covered and with sea level rise and be permanently submerged. Copyright: J M Matés Luque.

Figure 2.8: Small wooden boat (Ondarroa II) abandoned further up in the estuary, away from the main canal, in the marshes. It is covered by the high tide after being partially excavated. The constant sediments covering the wreck will make it disappear at some point in the future. Copyright: J M Matés Luque.

Coastal erosion

Coastal erosion is a serious issue that has intensified with climate change. In the following decades, it is expected that sea-level rise along with increase in storminess will erode coastal and estuarine zones at fast rates. Archaeological sites and heritage assets in these areas, especially on riverbanks, cliffs and beaches, are at risk of being lost in these circumstances and would require careful monitoring. Additionally, coastal erosion might expose new archaeological sites, previously buried in thick sediments. In Spain coastal erosion is affecting archaeological sites, such as those in the Arosa island, in Galicia (Mañana-Borrazás *et al.* 2020) and on the Asturias coast on the north of Spain where coastal erosion exposed remains of human bones, timbers and stone structures of the 19th century (Alonso Rodríguez and Matés Luque 2020: 297–299).

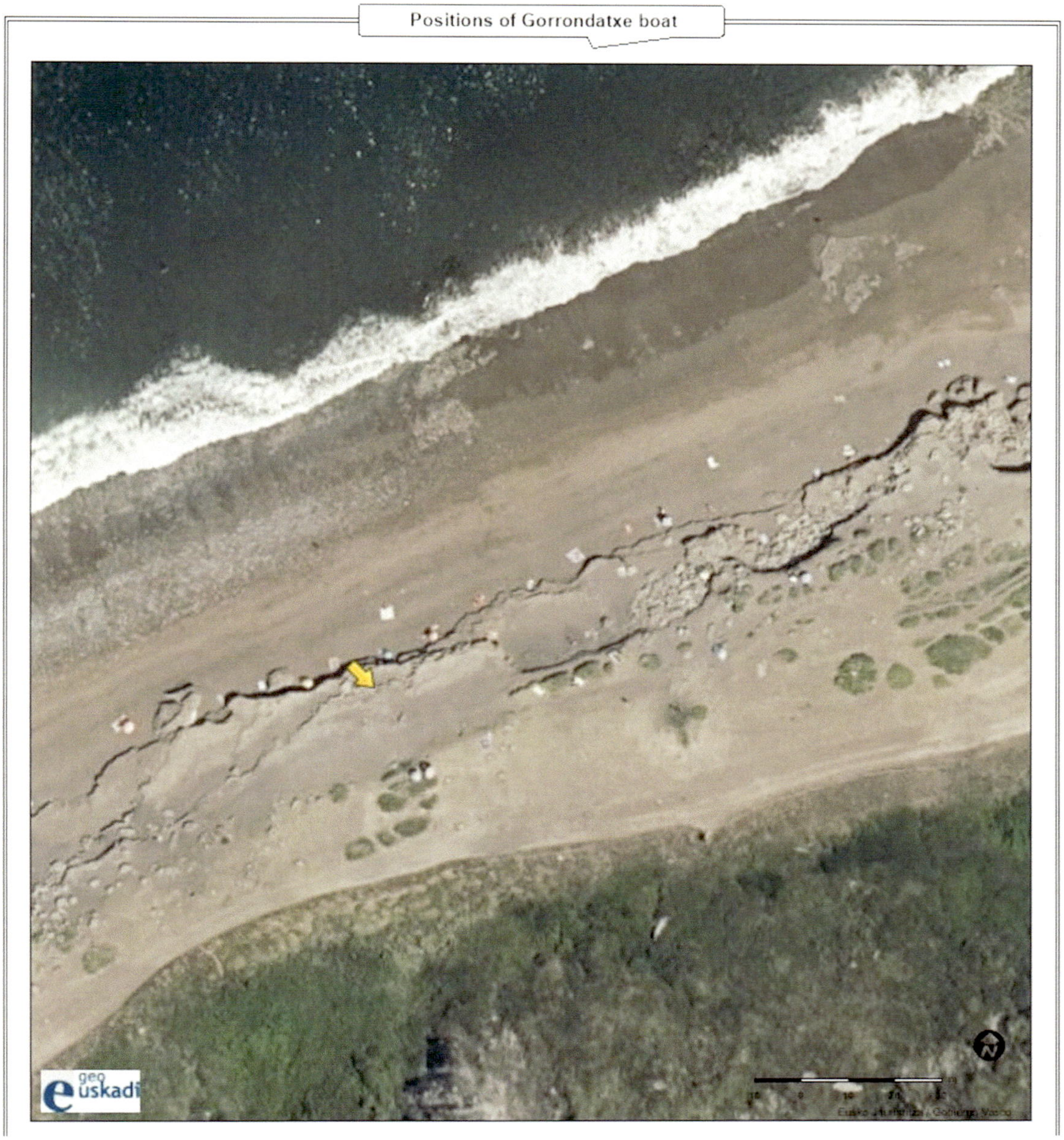

Figure 2.9: Position of the Gorrondatxe boat next to the border of the beach, behind the cliff. The sea is at low tide. Copyright: Geovisor Euskadi, modified by J M Matés Luque.

In the Oka River estuary examined for this study, remains of what was later identified as two mid-19th-century shipwrecks, were exposed when the sand which covered them was shifted by erosion and the river current (Gómez Bravo and Zallo Uskola 2014; Zallo Uskola and Gómez Bravo 2016). Since then, yearly monitoring of the sites has allowed assessing their conditions and level of coverage by sand providing a local baseline to measure and better understand coastal erosion in the area (Zallo Uskola and Gómez Bravo 2012; Gómez Bravo and Zallo Uskola 2021).

More recently, a clinker-built boat was exposed by coastal erosion on a nearly two-meter high cliff edge, by the sandy beach of Gorrondatxe in Getxo (Figure 2.9-2.11), known for preserving debris from iron furnaces that people used to dump from factories in the open sea but were later pushed to the coast by sea currents (Astibia 2012; Matés Luque 2023; Pujalte *et al.* 2015). In this study, and throughout the excavation of the boat (Figure 2.12), we have been monitoring regularly the distance of the clinker-built boat from the edge of the eroding cliff and the height of the cliff (Figure 2.13), as a means to assess the changes that are occurring at Gorrondatxe beach and the rhythm of coastal erosion in the area.

Figure 2.10: View of the Gorrondatxe boat from above the hill, looking down to the boat by the cliff. The sea is at low tide. Copyright: M Matés Luque.

Figure 2.11: View of the of the Gorrondatxe boat partially excavated. The sea and the edge of the cliff can be seen. Copyright: J M Matés Luque.

Figure 2.12: Excavation of the of the Gorrondatxe boat at the edge of the cliff by the sea (to the right). Erosion of the top layers of the beach can be seen in the middle of the photograph. Copyright: J M Matés Luque.

Figure 2.13: View of the height of the cliff, on top of which the of the Gorrondatxe boat is buried. Erosion by sea and wind can be observed. Copyright: J M Matés Luque.

Sea and ocean warming

From early on in climate change maritime archaeological research, the warming of the seas and oceans has been identified as a serious threat to wooden shipwrecks worldwide (e.g., González-Duarte *et al.* 2018; Wright 2016). Even though the impacts of this effect on wooden maritime heritage is not yet monitored in the Basque Country, we believe that there is a big opportunity to select wreck sites, such as the wooden boat remains seen in Figure 2.14, as monitoring stations where water temperature changes can be tracked along with the impacts on the wooden material remains. With frequent field surveys, site temperature measuring gauges and use of satellite imagery, it would be possible to monitor the water temperature changes in different seasons, and the evolution of seasonal submersion of the sites, particularly in relation to pluvial, fluvial and tidal flooding, plotting thus the influence of warming seas and observing the degradation of the material remains. All the above data could assist in preparing guidelines on safeguarding measures that might be required for intertidal archaeological sites at risk, and particularly for the protection of wooden shipwrecks. Moreover, analyses of the collected datasets could provide recommendations on potential impacts on other contemporary material culture of the intertidal zone, such as house foundations, quays, parks, recreational spaces, and agricultural lands that might become affected by phenomena related to warming sea temperatures.

Figure 2.14: Picture of the Mendexa I wreck, which has suffered a lot of damage over the years. In 2008 some parts of the wreck's sides were still upright. Unfortunately, they eventually collapsed. This boat is very easy to reach and can be used as a benchmark to monitor various factors of deterioration, including water temperature. Copyright: J M Matés Luque.

Shipworm rise due to warmer waters

Due to rising sea temperatures, shipworm numbers have been increasing, particularly in shipwreck sites, affecting the preservation levels of wooden remains (Gjelstrup Björdal and Gregory 2012; Palma and Santhakumaran 2014). Abandoned timber boats, such as the remains in Figure 2.15 in the intertidal zone, can become monitoring sites to identify the presence and quantity of such shipworm species, offering, again, a baseline from where it will be possible to assess locally levels of risk for timber remains in archaeological sites, but also in other contemporary usable structures (e.g., piers, staircases, quays, etc.).

Figure 2.15: Remains of one or two boats (research ongoing). They are fully covered at high tide, resulting to shipworms reaching them easily. The remains can be used for shipworm monitoring, as well as checks for other chemical and physical elements in the water, especially due to their proximity to a sewage collector. Copyright: J M Matés Luque.

pH levels

Recording and monitoring pH levels on metal wrecks due to increased ocean acidification caused by climate change has been noted as an important issue by many scholars around the world (e.g., Dunkley 2015: 222–225; Gregory *et al.* 2022: 8–10.). However, in the Basque Country, there are currently no provisions or processes to carry out such research. This relates largely to the disconnect between archaeology practitioners and scientists or stakeholders who monitor and manage environmental data. A few underwater metal wrecks are known to the author (Matés Luque and Armendariz 2018), but it has been observed that such sites are not necessarily considered as significant heritage assets by the local authorities and heritage stakeholders, due to their more recent date. Hence, there has not been more extensive research and monitoring. Despite this attitude towards post-medieval heritage of this type, knowing and monitoring the impacts of ocean acidification on metal wrecks is a very significant issue, specifically because recent research has highlighted the polluting effects that certain types of degradation can have on the natural ecosystems (Brennan 2024). Given the limited amount of data available in the case of the Basque Country, this research has identified that there is at least one iron barge (Figures 2.16-2.17), which was unfortunately affected by the opening of the Deusto canal (Matés Luque 2019a), that could be used with little effort for this type of research and monitoring of pH levels in the intertidal zone.

Figure 2.16: View of an old iron barge in the estuary at Bilbao. Only the bottom part of the boat is preserved. This type of iron is identifiable in old photographs from the 19th and 20th centuries showing the industrial activity of this estuary. Copyright: J M Matés Luque.

Figure 2.17: The dredging of the estuary that took place during the opening of the Deusto canal removed part of the riverbank where the barge is standing. Following that (sometime after this picture was taken) one end of the barge collapsed due to its unsupported weight. Copyright: J M Matés Luque.

Flora, fauna and biodiversity

Monitoring numbers and types of flora and fauna found by intertidal heritage remains can help obtain significant data on the impacts of climate change on the local ecosystems, as well as the levels of risk to an intertidal or underwater archaeological site from invasive species. More specifically for the Basque Country, monitoring changes (expansion or reduction) of seagrass (*Zostera noltii*), an endangered marine plant present in the Lea estuary, can provide insightful information on the local patterns and changes of the climate (Kersting 2016: 85). In fact, there is a wreck in the middle of the canal (Lekeitio X) in the Lea estuary whose recording was not authorised by the local environmental agency due to the presence of such seagrass (Figure 2.18). However, monitoring of the material remains along with the seagrass, known to be offering protective conditions for organic material remains, is possible and could offer data for safeguarding both cultural and natural heritage at this area (Krause-Jensen *et al.* 2019; UN SDG Target 11.4). Similarly, the presence or absence of other marine life including fish, birds, reptiles or insects living in the intertidal zones can provide opportunities for heritage monitoring and safeguarding. The long-polluted estuary of Bilbao has been becoming cleaner since the 1980s through efforts that seek a more environmentally friendly estuary with many factory closures as well as sewage recycling plants and therefore reduction of the polluted waste dumped into the water. Thus, in recent years water quality has improved bringing back many animals to the estuary (Franco San Sebastián 2020), including even seahorses (Uriarte 14 January 2017) which are very sensitive to water quality. An overlap of research and monitoring on marine life and maritime heritage would allow the collection of interdisciplinary data that could be used by physical scientists, biologists, ecologists and heritage professionals to plan and protect both natural and cultural heritage simultaneously.

Figure 2.18: Wreck covered by seagrass in the middle of the Lea estuary. The Environmental Agency did not authorise the excavation of this wreck. However, it authorised the survey of the estuary (when this picture was taken). Copyright: J M Matés Luque.

Conclusion

Climate change in the Basque Country is a very significant issue, currently handled by many different bodies, research centres, organisations, and governments (local, regional, autonomous and national). So far, most efforts have focused on how climate change will affect citizens, the natural environment and the landscape, while heritage and archaeology have not been considered as relevant. As presented in this paper, climate change effects are posing significant risks to heritage assets of the intertidal zone. At the same time intertidal archaeological sites and in particular shipwreck remains offer unique opportunities for coupled cultural and natural environment monitoring within fixed settings, where elements such as sea level and water temperature, coastal erosion, pH and presence or absence of different organisms can be measured. This approach can offer local climate data and place-based insights on the impacts of climate change on the material culture of the region. Additionally, wrecks and other archaeological structures could be utilised for wider conservation and rewilding efforts. This paper is an invitation to explore the possibilities of how intertidal archaeological sites could be better integrated in the efforts to tackle climate change and environmental deterioration in the region. By joining forces with other disciplines and applying interdisciplinary approaches to development or environmental projects, as well as site monitoring, it will be possible to safeguard heritage at risk, while gathering invaluable localised environmental and climate change data.

Acknowledgements

The ideas expressed in this article are based on fieldwork carried out in the last 10 years. The work started with our own resources and later funding was received from different organisations which need to be thanked and recognized. Without their support, the recording of the wrecks and structures would not have been done or would not have reached as far as it rests nowadays. Thus, thanks are due to the Heritage Centre of the Cultural Department of the Basque Country and to the Archaeological Service of the Bizkaia Regional Council for being the first bodies which supported our surveys. The Institute of Nautical Archaeology (INA) also funded this modest project. But a project like this, which involved the excavation of many wrecks, could not have been done without many friends and colleagues, who I would like to thank: Esther Alonso, David Alonso Vega, Xabier Armendariz, Ander González Alonso, Aitor Leniz Atxabal, Carlos Alejandro Machín Santaolalla, Judith Mineto, Daniel Miguel Nieva Sanz, Patricia Rodriguez, Jesús Serrano, Ander Silvano, Guillermo Velasco, and Tania Vicario Valencia. The author assumes sole responsibility for the information expressed in this article.

References

Alonso Rodríguez, N. and J. M. Matés Luque 2020. Arqueología en la frontera (del mar). Nuevas investigaciones sobre arqueología marítima en Bizkaia y Asturias, in I. Grau Sologestoa and J. A. Quirós Castillo (eds) *Arqueología de la Edad Moderna en el País Vasco y su entorno*: 290–305. Oxford: Archaeopress.

Astibia, H. 2012. Tunelboka y Gorrondatxe (Getxo, Bizkaia), fósiles humanos para el Antropoceno. Euskonews 640 (2012-09-28 / 2012-10-05), viewed 27 January 2020, <http://www.euskonews.eus/zbk/640/tunelboka-y-gorrondatxe-getxo-bizkaia-fosiles-humanos-para-el-antropoceno/ar-0640001002C/>.

Baika, K. 2012. Submerged archaeological remains at Cape Sounion, Attica, and implications for the study of relative sea-level changes since the Classical period in the central Aegean, in J. Henderson (ed.) *IKUWA 3: Beyond boundaries. The 3rd International Congress on Underwater Archaeology. London 2008*: 277–285. Bonn and Portsmouth: Deutsches Archäologisches Institut-Römisch-Germanische Kommission, Dr. Rudolf Habelt, Nautical Archaeology Society.

Berganza Gochi, Eduardo and J. L. Arribas Pastor 2014. *La Cueva de Santa Catalina (Lekeitio, Bizkaia): La intervención arqueológica. Restos vegetales, animales y humanos. Kobie. Bai 4*. Bilbao: Diputación Foral de Bizkaia.

Brennan, M.L. 2024. *Threats to Our Ocean Heritage: Potentially Polluting Wrecks*. Cham: Springer. <https://doi.org/10.1007/978-3-031-57960-8>

Campos Rodrigues, L. M, I. Puig Ventosa and S. Sastre Sanz 2017. *Adaptación al cambio climático de zonas urbanas costeras con elevada densidad de población e interés turístico y cultural en España*. Vilanova i la Geltrú: Fundación ENT.

Chust, G., A. Borja, A. del Campo, P. Liria, J. Franco, I. Muxika, R. Roa, J. G. Rodríguez and J. M. Garmendia 2013. *Inundabilidad de los estuarios y costa de Bizkaia por ascenso del nivel del mar mediante LiDAR topográfico y batimétrico: cartografía y evaluación de impactos*. Informe inédito elaborado por AZTI-Tecnalia para la Agencia Vasca del Agua - Uraren Euskal Agentzia.

Climate Central, Coastal Risk Screening Tool, viewed 15 February 2025, <https://coastal.climatecentral.org/map/13/-2.4778/43.3245/?theme=sea_level_rise&map_type=ice_sheet&basemap=roadmap&contiguous=true&elevation_model=best_available&ice_loss_level=1.0&ice_sheet=antarctic&refresh=true&water_unit=m>.

Dunkley, M. 2015. 'Climate is what we expect, weather is what we get' – Managing the potential effects of oceanic climate change on underwater cultural heritage, in W. J. H. Willems and H. P. J. van Schaik (eds) *Water and Heritage. Material, conceptual and spiritual connections*: 217–229. Leiden: Sidestone Press.

Franco San Sebastián, J., 27 May 2020, La recuperación de la fauna en la ría de Bilbao, Azti Newsletter, viewed 7 April 2024, <https://www.azti.es/la-recuperacion-de-la-fauna-en-la-ria-de-bilbao/>.

Gjelstrup Björdal, C. and D. Gregory 2012. *WreckProtect: Decay and protection of archaeological wooden shipwrecks*. Oxford: Archaeopress.

Gobierno Vasco. 2015. *Estrategia de Cambio Climático 2050 del País Vasco*. Vitoria-Gasteiz: Servicio Central de Publicaciones del Gobierno Vasco.

Gobierno Vasco. 2022. *Plan estratégico Cultura 2028*. Vitoria-Gasteiz: Departamento de Cultura y Política Lingüística, Gobierno Vasco.

Gómez Bravo, M., and J. C. Zallo Uskola 2014. Nuevos hallazgos en la ría de Gernika-Mundaka (Bizkaia): los pecios Laida I y Laida II, in X. Nieto Prieto and M. Bethencourt Núnez (eds) *Arqueología subacuática española. Actas del I Congreso de Arqueología Náutica y Subacuática Española. Cartagena, 14, 15 y 16 de marzo de 2013, II*: 133–141. Cádiz: Servicio de Publicaciones de la Universidad de Cádiz.

Gómez Bravo, M. and J. C. Zallo Uskola 2021. Pecios Laida I, II, Txatxarramendi y Porturas e inspección de arenas (Ibarrangelu / Sukarrieta). *Arkeoikuska* 2020: 263–266.

Gómez Royuela, M. 2016. *Impactos, vulnerabilidad y adaptación al cambio climático en el sector turístico*. Madrid: Ministerio de Agricultura, Alimentación y Medio Ambiente Secretaría General Técnica Centro de Publicaciones.

González-Duarte, M.M., T. Fernández-Montblanc, M. Bethencourt and A. Izquierdo 2018. Effects of substrata and environmental conditions on ecological succession on historic shipwrecks. *Estuarine, Coastal and Shelf* Science 200: 301–310. <https://doi.org/10.1016/j.ecss.2017.11.014>

Gregory, D., T. Dawson, D. Elkin, H. Van Tilburg, C. Underwood, V. Richards, A. Viduka, K. Westley, J. Wright, and J. Hollesen 2022. Of Time and Tide: The Complex Impacts of Climate Change on Coastal and Underwater Cultural Heritage. *Antiquity* 96.390: 1396–1411. <https://doi.org/10.15184/aqy.2022.115>

Hafner, A. and C. J. Underwood 2020. Introduction to the impact of climate change on underwater cultural heritage and the decade of ocean science for sustainable development 2021–2030, in A. Hafner, H. Öniz, L. Semaan and C. J. Underwood (eds) *Heritage at risk. Special Edition. Heritage under water at risk: Threats, Challenges, Solutions*: 118–126. Paris: International Council on Monuments and Sites.

Ibisate González de Matauco, A., A. Ollero Ojeda and O. Ormaetxea Arenaza. 2000. Las inundaciones en la vertiente cantábrica del País Vasco en los últimos veinte años: principales eventos, consecuencias territoriales y sistemas de prevención. *Serie Geográfica* 9: 177–186.

Ihobe (Ingurumen Jarduketarako Sozietate Publikoa/ Sociedad Pública de Gestión Ambiental) 2019. *Evaluación de la vulnerabilidad y riesgo de los municipios vascos ante el cambio climático.* Bilbao: Ihobe, Sociedad Pública de Gestión Ambiental Departamento de Medio Ambiente, Planificación Territorial y Vivienda Gobierno Vasco.

Izaguirre, M. and L. Valdés. 1998. Avance de excavación del pecio del siglo XV de Urbieta (Gernika). *Itsas Memoria. Revista de Estudios Marítimos del País Vasco* 2: 35–41.

Izaguirre, M., L. Valdés and J. M. Matés. 1999. Avance de excavación del Pecio del s. XV de Urbieta (Gernika). *Arkeoikuska* 1998: 392–398.

Izaguirre, M., L. Valdés and J. M. Matés Luque and I. Pujana. 2001. State of the excavation works of the 15th century shipwreck in Urbieta (Gernika, Spain), in F. Alves (ed.) *Trabalhos de Arqueologia. Proceedings. International Symposium on Archaeology of Medieval and Modern Ships of Iberian-Atlantic Tradition. Hull remains, manuscripts and ethnographic sources* 18: 449–454. Lisboa: Instituto Português de Arqueologia.

Kersting, D. K. 2016. *Cambio climático en el medio marino español: impactos, vulnerabilidad y adaptación.* Madrid: Oficina Española de Cambio Climático, Ministerio de Agricultura, Alimentación y Medio Ambiente.

Kolaiti, E. and N. Mourtzas 2020. New Insights on the Relative Sea Level Changes during the Late Holocene along the Coast of Paros Island and the Northern Cyclades (Greece). *Annals of Geophysics* 63.6. <https://doi.org/10.4401/ag-8504>

Krause-Jensen, D., O. Serrano, E. T. Apostolaki, D. J. Gregory and C. M. Duarte 2019. Seagrass sedimentary deposits as security vaults and time capsules of the human past. *Ambio* 48: 325–335.

Losada Rodríguez, I. J., C. Izaguirre Lasa and P. Díaz Sismal 2014. *Cambio climático en la costa española.* Madrid: Ministerio de Agricultura, Alimentación y Medio Ambiente. Gobierno de España.

Lubao, C. B. and E. Ichumbaki 2023. Fishing songs from Kilwa Kisiwani, Tanzania: a case study of intangible marine cultural heritage on the Swahili coast. *Journal of Maritime Archaeology* 18: 165-195. <https://doi.org/10.1007/s11457-023-09356-5>

Mansilla Hortigüela, R. 2010. Molino de marea de Marierrota o Isuntza. *Arkeoikuska* 2009, 286–289

Mañana-Borrazás, P., R. Blanco-Chao, M. J. Bóveda-Fernández D. Cajade-Pascual, M. Costa-Casais, A. Güimil-Fariña, E. López-Romero, S. Vázquez-Collazo, and X. Ignacio Vilaseco-Vázquez 2020. Lo que nos cuenta la marea. Prehistoria en el islote de Guidoiro Areoso (A Illa de Arousa, Galicia) a la luz de las últimas intervenciones, in A. Carretero Pérez and C. Papí Rodes (eds) *Actualidad de la Investigación arqueológica en España I (2018-2019). Conferencias impartidas en el Museo Arqueológico Nacional*: 159–176. Madrid: Secretaría General Técnica. Subdirección General de Atención al Ciudadano, Documentación y Publicaciones.

Martínez Sarracina, P. 2023. 40 años de la gran riada: el recuerdo de las horas más oscuras. *El Correo August 2023* viewed 26 April 2024, <https://www.elcorreo.com/bizkaia/inundaciones-bilbao-1983-40-anos-gran-riada-recuerdo-horas-oscuras-20230825122657-nt.html>.

Matés Luque, J. M. 2013. Muelles. *Arkeoikuska* 2012: 146–152.

Matés Luque, J. M. 2014. Muelle de Arriaga. *Arkeoikuska* 2013: 194–196.

Matés Luque, J. M. 2015. Estructuras de pesca de la ría de Bilbao, entre San Antón y La Peña. *Arkeoikuska* 2014: 196–198.

Matés Luque, J. M. 2016a. Maritime archaeology as a tool to understand a river town: the example of Bilbao (Basque Country, Spain), in R. Castillo Belinchón, A. Ramírez Pernía and L. A. Torres Sobrino (eds) *Actas del V Congreso Internacional de Arqueología Subacuática. IKUWA V. Un patrimonio para la humanidad Cartagena, 15-18 October 2014*: 781–790. Madrid: Secretaría General Técnica. Subdirección General de Documentación y Publicaciones.

Matés Luque, J. M. 2016b. Malecón de Lazunarri. *Arkeoikuska* 2015: 236–243.

Matés Luque, J. M. 2018a. Arqueología marítima en la ría de Bilbao: cuando los árboles dejan ver el bosque, in I. Vivas Ziarrusta and A. Lekerikabeaskoa Gaztañaga (eds) *La Ribera de Deusto y Zorrotzaurre en la Ría de Bilbao. Análisis urbano, sociocultural y estético del espacio marítimo. Maritimidades* 5: 299-328. Bilbao: Museo Marítimo Ría de Bilbao. Bilboko Itsasadarra Itsas Museoa.

Matés Luque, J. M. 2018b. Cable de telecomunicaciones entre USA y la playa de Arrietara (proyecto Marea). *Arkeoikuska* 2017: 316–317.

Matés Luque, J. M. 2019a. Apertura del Canal de Deusto. *Arkeoikuska* 2018: 284–288.

Matés Luque, J. M. 2019b. Arqueología intermareal en Bizkaia: documentando la flota abandonada de embarcaciones vascas en los estuarios del Lea y del Artibai. Problemas y soluciones. *Kobie Paleoantropología* 37: 47–60.

Matés Luque, J. M. 2020a. Arqueología intermareal en Bizkaia. Documentación de embarcaciones de madera en los estuarios de Lea (Lekeitio y Mendexa) y de Artibai (Ondarroa). *Arkeoikuska* 2019: 382–386.

Matés Luque, J. M. 2020b. Maritime archaeology in Biscay, Basque Country. Facts, acts, research and opportunities, in N. Raad and C. Cabrera Tejedor (eds) *Ships, Boats, Ports, Trade, and War in the Mediterranean and Beyond; Proceedings of the Maritime Archaeology Graduate Symposium 2018, Oxford, 3-4 February*: 127–142. Oxford: BAR Publishing.

Matés Luque, J. M. 2020c. When the tide is low. Intertidal archaeology in the estuaries of the province of Bizkaia (Basque Country, Spain). *Quaterly International* 566-567: 315–322. <https://doi.org/10.1016/j.quaint.2020.05.008>

Matés Luque, J. M. 2022a. Cable submarino Grace Hopper entre USA y Sopela en la playa de Arrietara. *Arkeoikuska* 2021: 314–316.

Matés Luque, J. M. 2022b. Embarcaciones en la zona intermareal de la ría de Lea-Artibai: Mendexa II, IV y Ondarroa I. *Arkeoikuska* 2021: 346-48.

Matés Luque, J. M. 2023. Junto al mar. Un pecio de madera a tingladillo abandonado en la playa de Gorrondatxe (Getxo, Bizkaia). *Kobie. Antropología Cultural* 40: 55–67.

Matés Luque, J. M. and X. Armendariz 2018. Arqueología subacuática en Bizkaia: 35 años después, in L. Agudo Pérez, C. Duarte, A. Garcia Escarzaga, J. M. Geiling, A. Higuero Pliego, S. Núñez de la Fuente, F. J. Rodríguez Santos and R. Suarez Revilla (eds) *Actas de las IX Jornadas de Jóvenes en Investigación Arqueológica, Santander 8-11 June. JIA 2016*: 533–542 Santander: Instituto Internacional de Investigaciones Prehistóricas de Cantabria.

NASA Sea Level Projection Tool., viewed 15 February 2025, <https://sealevel.nasa.gov/ipcc-ar6-sea-level-projection-tool>.

Palma, P. and L. N. Santhakumaran 2014. *Shipwrecks and global 'worming'*. Oxford: Archaeopress.

Peev, P., R. H. Farr, V. Slavchev, M. J. Grant, J. Adams and G. Bailey 2020. Bulgaria: Sea-Level Change and Submerged Settlements on the Black Sea, in G. Bailey, N. Galanidou, H. Peeters, H. Jöns and M. Mennenga (eds) *The Archaeology of Europe's Drowned Landscapes*: 393–412. Cham: Coastal Research Library, vol 35. Springer.

Pérez-Álvaro, E. 2016. Climate change and underwater cultural heritage: Impacts and challenges. *Journal of Cultural Heritage* 21: 842–848.

Perez-Alvaro, E., and R. Boswell 2025. Integral oceans heritage of indigenous communities: Its value for good health and well-being. *Social Sciences & Humanities Open* 11: 101245. <https://doi.org/10.1016/j.ssaho.2024.101245>

Pujalte, V., H. Astibia, F. Aizpiri and A. Payros. 2015. Las playas cementadas del "Antropoceno" de Bizkaia, País Vasco: origen y degradación. *Geogaceta* 57: 123–126.

Pujana, I. and A. Díez. 1999 Canal de la ría (Gernika). *Arkeoikuska* 1998: 376–381.

Sainz de Murieta, E., L. M. Abadie and I. Galarraga. 2018. Aumento del nivel del mar en la costa vasca: una aproximación probabilística. *Ekonomiaz* 94.2: 324–44.

Tellería Sarriegi, E. 2011. *Memoria de actividades arqueológicas. Molino de Marierrota, Mendexa (Bizkaia).* Vitoria-Gasteiz: Ondare Babesa.

Tellería Sarriegi, E. 2012. Molino de Marierrota. *Arkeoikuska* 2011: 273–274.

The Paris Agreement, UN Climate Action, viewed 15 February 2025, <https://www.un.org/en/climatechange/paris-agreement>.

United Nations (UN) Sustainable Development Goals (SDGs), viewed 15 February 2025, <https://sdgs.un.org/goals>

Uriarte, J., January 2017, Cabalgando por la ría, El Correo, viewed 18 April 2024, <https://www.elcorreo.com/bizkaia/201701/14/cabalgando-20170113191721.html>.

Velentza, K. 2023. Maritime Archaeological Research, Sustainability, and Climate Resilience. *European Journal of Archaeology* 26.3: 359–377. <https://doi.org/10.1017/eaa.2022.48>

Velentza, K. 2024. Traditional Shipbuilding on the Island of Samos, Greece: Recording the Tangible and Intangible Data. *International Journal of Nautical Archaeology*. <https://doi.org/10.1080/10572414.2024.2342281>

Wright, J. 2016. Maritime Archaeology and Climate Change: An Invitation. *Journal of Maritime Archaeology* 11: 255–270. <https://doi.org/10.1007/s11457-016-9164-5>

Zallo Uskola, J. C and M Gómez Bravo 2012. Evaluación arqueológica de los pecios Laida I y Laida II (Ría de Mundaka, Bizkaia). *Kobie Paleoantropología* 31: 207–226.

Zallo Uskola, J. C and M Gómez Bravo 2016. Nuevas embarcaciones a tingladillo en la ría de Gernika (Bizkaia), in R. Castillo Belinchón, A. Ramírez Pernía and L. A. Torres Sobrino (eds) *Actas del V Congreso Internacional de Arqueología Subacuática. IKUWA V. Un patrimonio para la humanidad Cartagena, 15-18 October 2014*: 757–768. Madrid: Secretaría General Técnica. Subdirección General de Documentación y Publicaciones.

3.
Tracing Climate Change Impacts on Maritime Heritage Resources in Lebanon Through Satellite Imagery

Sirine Saad Eldine Ghiye

This paper explores a method that could predict the impact of climate change on coastal archaeological sites in northern Lebanon, with the use of satellite imagery. The case study examined is the area around the coastal town of Chekka. The archaeological potential of Chekka is high due to its natural basin that is thought to be an ancient harbour. This study focusses on changes in sea surface temperature (SST) from 1985 to 2024, the sea surface salinity (SSS) recorded in the last two decades, and changes to the coastline from 1985 to 2023. The analysis of these sea characteristics is carried out through research of satellite imagery primarily from Landsat 8, equipped with two bands of the thermal infrared sensor (TIRS), Landsat 7, which carries the Enhanced Thematic Mapper Plus (ETM+) sensor, and Landsat 5 TM. After calculating SST in QGIS, an increase of 1° Celsius in sea surface temperature over the past two decades has been noted. This observation highlights the already noticeable impacts of climate change and the potential repercussions and future vulnerabilities that this situation creates for the coastal cultural heritage in the Chekka region.

Introduction

Climate change is one of the most pressing challenges that humanity is currently facing (Lange 2020). One of the clearest indicators of climate change is the increase in carbon dioxide (CO2) concentrations in the Earth's atmosphere, also known as atmospheric CO2. Recognising variations in CO2 concentrations requires data over extended periods, which can be compared to baseline data that pre-dates the industrial revolution (Kennedy *et al.* 2002).

The sea surface is a dynamic and complex environment that is influenced by various factors, including wind, temperature, salinity, currents, and atmospheric conditions (Global Climate Observation System n.d.). The increase in atmospheric CO2 has led to significant changes in sea surface temperature (SST) (Spalding 2011). Existing studies have shown that detecting changes in SST can be used to monitor climate change variations and their impact on marine environments, as well as for modelling sea-level rise (e.g., Abou Dagher *et al.* 2012).

Tracking SST involves a combination of measurement techniques and technologies that enable analysis of temperature shifts over specific periods. Advancements in satellite technology that enable SST measurements have greatly expanded our ability to monitor and understand global ocean temperature patterns, leading to a more comprehensive understanding of the Earth's oceans and their dynamic behaviours (Bourouhou and Salmun 2021). For example, Kokal *et al.* (2022) examined SST anomalies in the Marmara Sea in Turkey over the past 32 years. Their study used various satellite imagery, such as Landsat-8 and Sentinel-3, where the sensors measured large areas of sea temperature via the thermal infrared region of the electromagnetic spectrum. The results indicated an increase in SST by 2°C over the last 32 years in the Marmara Sea.

The Future by the Sea (Archaeopress 2026): 56–69

Climate change also impacts ocean circulation, which affects marine biodiversity and ocean nutrient distribution and can lead to negative impacts on maritime cultural heritage (Ghiye 2023). For example, studies have shown that wood-eating organisms, like the shipworm species *Teredo navalis* and *Lyrodus pedicellatus*, thrive in warmer waters (e.g., Gregory *et al.* 2022).

This paper presents a methodology that can help measure changes in SST and sea surface salinity (SSS) in the Chekka area of northern Lebanon over the past 30 years (Figure 3.1) with the aim to highlight how human-induced global warming can lead to significant impacts on cultural heritage.

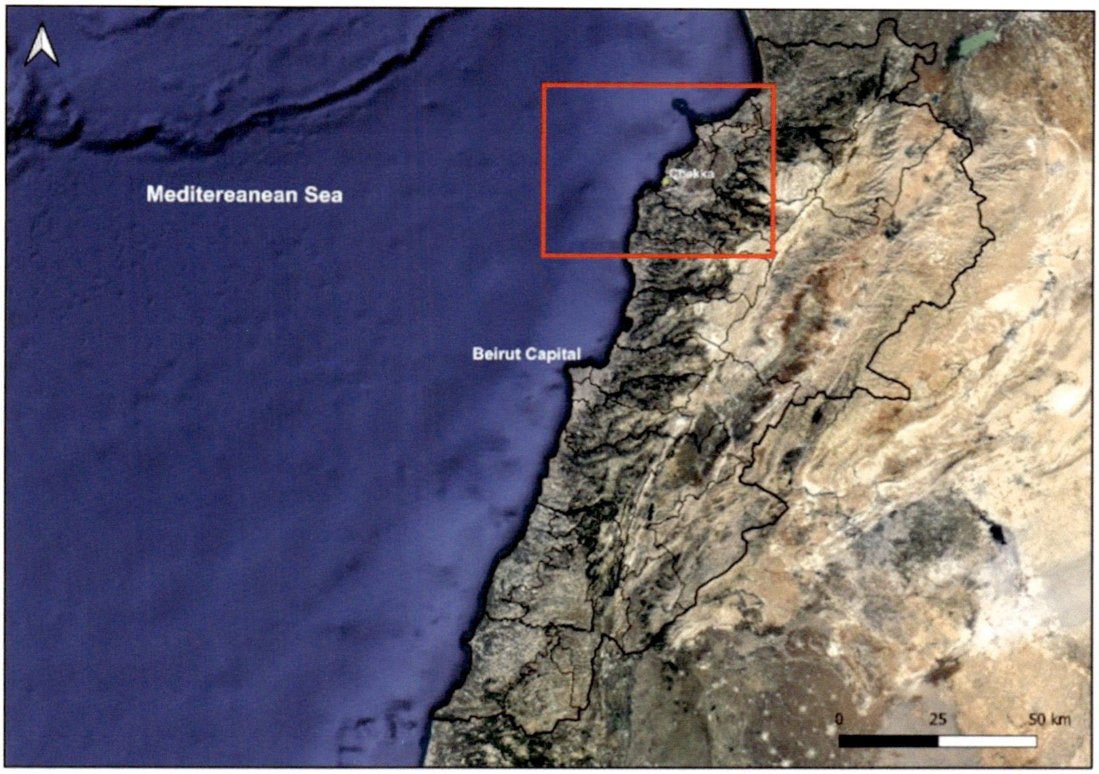

Figure 3.1: Map of Lebanon showing the location of Chekka, the case study of this project. Copyright: Sirine Saad Eldine Ghiye.

Chekka has a long history of human occupation, starting probably from the Bronze Age especially if the site is the same as Šigata/Chikitta mentioned in the Amarna letters from the 14th century BC. Geographically, Chekka has two potential locations for ancient harbours, one to the north and one to the south of the headland of Chekka (El Safadi 2014). An archaeological survey there in 1974 documented several maritime-related artefacts, including anchors and ballast mounds, indicating past maritime activities and the potential presence of further archaeological remains in the area.

Before the early 20th century, Chekka was covered by coastal salt marshes and agricultural land for olive production. This changed after a cement factory was constructed on the Bay of Chekka in 1931, and a second factory built two decades later. The construction of the factories changed the nature of the shoreline, particularly because of an artificial jetty which extends 750m into the sea (Abou Dagher *et al.* 2012). Erosion was recorded as far as 81m inland between 1970 and 1994, with accretion around the jetty reaching a maximum of 94m (Abou Dagher *et al.* 2012). Over the same period, the coastline became densely populated, and many agricultural areas vanished.

Archival records, historical information and geomorphological factors suggest that Chekka and its surrounding region have a high potential for the preservation of terrestrial and maritime archaeological remains, a lot of which might be currently undocumented. Hence, a study of sea surface temperature, sea salinity, and changes in coastlines in a specific area of the Lebanese coastlines can offer valuable insights and a first assessment of the potential impacts of climate change on known and unknown heritage assets on the coast or under water. The outcomes can guide conservation efforts, support adaptive management, and serve as a foundation for broader regional studies

Aims and methodology

The study has two aims. Firstly, to use satellite imagery to calculate the increase of SST and the salinity of water in Chekka, in northern Lebanon, using the methodology outlined below. Secondly, to assess the potential impact of underwater and coastal archaeological sites in the area by the recorded changes in the marine environment due to climate change.

Remote sensing is a common method used for maritime heritage research, including use of satellite imagery and marine geophysics (e.g., Yang *et al.* 2013). More specifically, in the last decade various studies have proven that satellite imagery can play a crucial role in the study of maritime archeological sites by providing valuable and efficient means of investigating the underwater and coastal environments. It is a way to identify and detect sites, but also to understand and record site formation processes and potential management interventions. Recently, satellites have been equipped with thermal sensors to measure radiance from the surfaces of land, seas and other bodies of water. This is crucial for understanding oceanic and climatic patterns, as well as monitoring environmental changes. Through satellite imagery, the impact of climate change is seen in the increases of surface temperatures of bodies of water and of sea salinity in oceans and seas.

For this paper, data and images from Landsat satellites have been used in an attempt to measure the variations in temperature and salinity of the sea around Chekka in northern Lebanon. The Landsat program is a series of Earth-observing satellite missions jointly operated by the US Geological Service (USGS) and NASA. The Landsat program has been providing a continuous record of the Earth's surface since the launch of the first satellite in 1972. The primary mission of Landsat is to collect multispectral imagery of the Earth's surface for various applications, including land use, environmental monitoring, and resource management (Bourouhou and Salmun 2021). Landsat satellites capture imagery in multiple spectral bands, including visible, near infrared, and thermal infrared. The sensors also capture data in multiple bands, allowing for multispectral analysis which is particularly useful for measuring environmental changes, including changes in sea surface temperature and salinity (Reul *et al.* 2020). The sources of data used in this study and are Landsat 5, 7 and 8 (Table 3.1), which provide a large amount of data that covers the land and coast of Lebanon.

Table 3.1: Dataset used for measuring SST and SSS. Copyright: Sirine Saad Eldine Ghiye.

Landsat ID	Row/Path	Dated	Season
Landsat 7 ETM+ L1TP_174036_19990722_20211204_02_T1	036/174	July 1999	Summer
Landsat 8 LC08_L1TP_174036_20240312_20240312_02_RT	036/174	January 2024	Winter
Landsat 8 LC08_L1TP_174036_20230801_20230805_02_T1	036/174	August 2023	Summer
Landsat 5 TM LT05_L1TP_174036_19850128_20200918_02_T1	036/174	January 1985	Winter
Landsat 8 LC09_L1TP_174036_20220211_20230428_02_T1	036/174	February 2022	Winter

Landsat 5 was active between 1984 and 2013. It carried a multispectral scanner system (MSS) and thematic mapper (TM). Landsat 7 was active from 1999 to 2024. The primary features of Landsat 7 included a panchromatic band with 15-meter spatial resolution, an onboard full aperture solar calibrator, five percent absolute radiometric calibration and thermal infrared sensors (TIRS) instruments (USGS n.d.). Landsat 8 has been active since 2013. The satellite consists of the Operational Land Imager (OLI) and TIRS. Through TIRS, land temperature in two thermal bands can be measured with a new technology that applies quantum physics to detect heat. The process works by measuring the amount of infrared light emitted from the earth and the radiant temperature of water bodies (Zhou *et al.* 2018). Both sensors cover the landmass with 30m of resolution on the visible near-infrared (NIR) and short-wave infrared (SWIR) spectra, 100 meters on the thermal spectrum and 15 meters on the panchromatic band.

This research utilises data from thermal infrared bands from Landsat 8 to calculate sea surface temperature (SST). Additionally, it involves data from the thermal mapper (TM) of Landsat 5, where Band 3 (visible red, with a wavelength range of 0.63 - 0.69μm, 30m resolution) and Band 4 (near infrared, with a wavelength range of 0.76 - 0.90μm, 30m resolution) are applied. The older Landsat 5 series help in recreating a historical assessment for Chekka, because Landsat 5 provided Earth imagery data for nearly 29 years, enabling, thus, access to images of Chekka dating back to 1985.

Sea Surface Temperature

The data from the Landsat program were processed in the geographic information system software QGIS. The coordinate system used is the WGS84 UTM zone 36N, ESPG: 32636 which covers Lebanon. The WGS84 datum provides a standardised reference for expressing locations on the Earth's surface, and the purpose of UTM is to provide a mapping system that minimises distortion within each zone. For the processing the following steps were followed:

1. The SST was measured on the raster calculator, following the four steps detailed by Kokal *et al.* (2022: 1). Subsequently, the raw image data from Landsat was converted to top of atmosphere spectral radiance, which is the amount of radiation coming from the area. For that the following equation was used:

$$L_\lambda = M_L Q_{cal} + A_L$$

Where L represents the radiance, M_L represents the band-specific multiplicative rescaling factor, Q_{cal} represents the quantized and calibrated standard product pixel values for Band 10, and A_L represents the band specific additive rescaling factor provide in the metadata profile (MLT) as shown in Figure 3.2.

```
    RADIANCE_MINIMUM_BAND_1 = -63.03941
    RADIANCE_MAXIMUM_BAND_2 = 782.32098
    RADIANCE_MINIMUM_BAND_2 = -64.60432
    RADIANCE_MAXIMUM_BAND_3 = 720.90210
    RADIANCE_MINIMUM_BAND_3 = -59.53233
    RADIANCE_MAXIMUM_BAND_4 = 607.90527
    RADIANCE_MINIMUM_BAND_4 = -50.20101
    RADIANCE_MAXIMUM_BAND_5 = 372.00769
    RADIANCE_MINIMUM_BAND_5 = -30.72052
    RADIANCE_MAXIMUM_BAND_6 = 92.51493
    RADIANCE_MINIMUM_BAND_6 = -7.63991
    RADIANCE_MAXIMUM_BAND_7 = 31.18248
    RADIANCE_MINIMUM_BAND_7 = -2.57506
    RADIANCE_MAXIMUM_BAND_8 = 687.98145
    RADIANCE_MINIMUM_BAND_8 = -56.81373
    RADIANCE_MAXIMUM_BAND_9 = 145.38904
    RADIANCE_MINIMUM_BAND_9 = -12.00627
    RADIANCE_MAXIMUM_BAND_10 = 22.00180
    RADIANCE_MINIMUM_BAND_10 = 0.10033
    RADIANCE_MAXIMUM_BAND_11 = 22.00180
    RADIANCE_MINIMUM_BAND_11 = 0.10033
  END_GROUP = MIN_MAX_RADIANCE
  GROUP = MIN_MAX_REFLECTANCE
    REFLECTANCE_MAXIMUM_BAND_1 = 1.210700
    REFLECTANCE_MINIMUM_BAND_1 = -0.099980
    REFLECTANCE_MAXIMUM_BAND_2 = 1.210700
    REFLECTANCE_MINIMUM_BAND_2 = -0.099980
    REFLECTANCE_MAXIMUM_BAND_3 = 1.210700
    REFLECTANCE_MINIMUM_BAND_3 = -0.099980
    REFLECTANCE_MAXIMUM_BAND_4 = 1.210700
    REFLECTANCE_MINIMUM_BAND_4 = -0.099980
    REFLECTANCE_MAXIMUM_BAND_5 = 1.210700
    REFLECTANCE_MINIMUM_BAND_5 = -0.099980
    REFLECTANCE_MAXIMUM_BAND_6 = 1.210700
    REFLECTANCE_MINIMUM_BAND_6 = -0.099980
```

Figure 3.2: MLT profile of Landsat image. Copyright: Sirine Saad Eldine Ghiye.

2. The top of atmosphere reflectance was calculated to provide the ratio of radiation reflected to the incident solar radiation on a given surface. The results depend on the property of the material itself. The reflectance is applied on the two OLI bands 4 and 5 for Landsat 8, and the process applied on band 6 (IR), band 4 (NIR) and band 3 (red visible) for Landsat 7. This is in order to use them for calculating the brightness temperature with the equation:

$$\rho\lambda = (M_\rho Q_{cal} + A_\rho)/\sin \Theta_{SE}$$

Where ρ_λ represents the raw bands to top-of-atmosphere reflectance, M_ρ provides a multiplicative rescaling factor, Q_{cal} represents quantized and calibrated standard product pixel values, A_ρ provides an additive rescaling factor, and Θ_{SE} is the local solar elevation angle.

3. The brightness temperature is the measurement of radiance of the microwave radiation traveling upward from the top of atmosphere to the satellites. It is calculated using the equation below where K_1 and K_2 provide a constant value for TIRS, and L is the radiance calculated in step 1. The result of the equation will return the temperature in Kelvin from degrees Celsius.

$$T_{brightness} = \frac{K_2}{Ln(\frac{K_1}{\lambda_1} + 1)} - 273.15$$

4. The process continues in progression to find the emissivity ε of each pixel in order to convert the brightness temperature into land surface temperature or sea surface temperature. The emissivity is calculated from the value of Normal Difference Vegetation Index (NDVI) which is provided by the visible red and near infrared bands. The method used here follows that of Advan and Jovanovska (2016), which is the same used for calculating land surface temperature. The result returns the maximum and

minimum values of NDVI which are used in finding the proportion of vegetation in reason to rescale the NDVI between values representing bare soil and total vegetation.

Sea Surface Salinity

After calculating the SST, it was also possible to explore changes of sea surface salinity (SSS) in the Chekka area, especially where the cement factories were established during the 20th century. This gave a first indication of how various anthropogenic activities have been affecting negatively sea morphology, components and geology.

Changes in sea surface salinity (SSS), or the Normalized Difference Salinity Index (NDSI), measures the relationship between SSS and SST due to climate change (e.g., Ansari and Akhoondzadeh 2019; Kim *et al.* 2023; Marghani and Hashim 2011). There are two methods for calculating the salinity of seawater: the direct *in-situ* method in which samples are collected from seas, oceans or rivers and analyzed in the laboratory, and the indirect method which depends on data collected through remote sensing, such as from the data of Landsat 8 and 5 used in the present study. The accuracy of the latter method to measure SSS from satellite images is highly dependent on the accuracy of surface reflectance obtained through atmospheric correction algorithms and the reliability of parameter retrieval algorithms (Muhsi *et al.* 2022). Through the microwaves radiometer operated at frequency 1.4 GHz, SSS data are available since 2010 with regular repeat global coverage. The calculation of brightness temperature and emissivity are related to the dielectric and geometric characteristics of the sea surface (Boutin *et al.* 2021). The main unit of measurement is parts per thousand (PPT), which refers to the quantity of salt dissolved into the sea (Reul *et al.* 2020), but practical salinity units (PSU) are also used. The use of data from the Landsat series was essential to find the changes of the last 20 years within the Chekka area and how this change is linked to climate change. The main equation used is:

$$NDSI = \frac{TM\ Band\ 2 - TM\ Band\ 5}{TM\ Band\ 2 + TM\ Band\ 5}$$

Where the Thermal Mapper (TM) Band 2 is representing the blue band in Landsat 8 with 30m resolution and 0.450-0.515μm wavelength, useful for bathymetric mapping and distinguishing soil and vegetation, while the Thermal Mapper (TM) Band 5 is the near infrared with 0.845-0.885μm wavelength that represent the biomass content and shorelines (Acharya and Yang 2015).

The methodology above enabled the processing of the satellite data in QGIS to generate thermal maps showing temperature and salinity quantities for each pixel. All of that was mapped on the coastline of Chekka, which was digitised and morphologically analysed for changes caused by sea level changes and storms.

Results

In this study, SST data for the winters of 1985 and 2024, as well as the summers of 1999 and 2023 were compared. Analysis of maps and temperature profiles illustrate changes in temperature along the coastline and land, particularly near the cement factories and areas with high population densities. These changes are attributed to factors such as toxic gas emissions, coastal development, and construction without adequate consideration for ecosystem preservation (Ofremu *et al.* 2025). Average salinity, which can lead to degradation of underwater cultural heritage assets, and changes to coastlines, as a result of erosion, were also explored.

Winter season

During the winter, January is the coldest and wettest month in Lebanon. Rainfall is common throughout the country, with coastal areas receiving moderate rainfall and mountainous regions experiencing heavier precipitation. In general, the average air temperature for coastal areas ranges between 10°C to 17°C. The SST and the coastal land temperatures of Chekka were documented through satellite imagery for January of 1985 and January of 2024 (Figures 3.3-3.4). In 2024, the temperature on the coastline was typically 17°C to 20°C, and substantially higher next to the cement factories. This is an increase from 1985, when temperatures did not exceed 15.2°C. This increase can also be seen in the change of the temperature profiles in graphs of Figure 3.5, where the maximum temperature in 2024 was 23°C compared to 18°C in 1985, and the average temperature increased from 14.7°C in 1985 to 19°C in 2024.

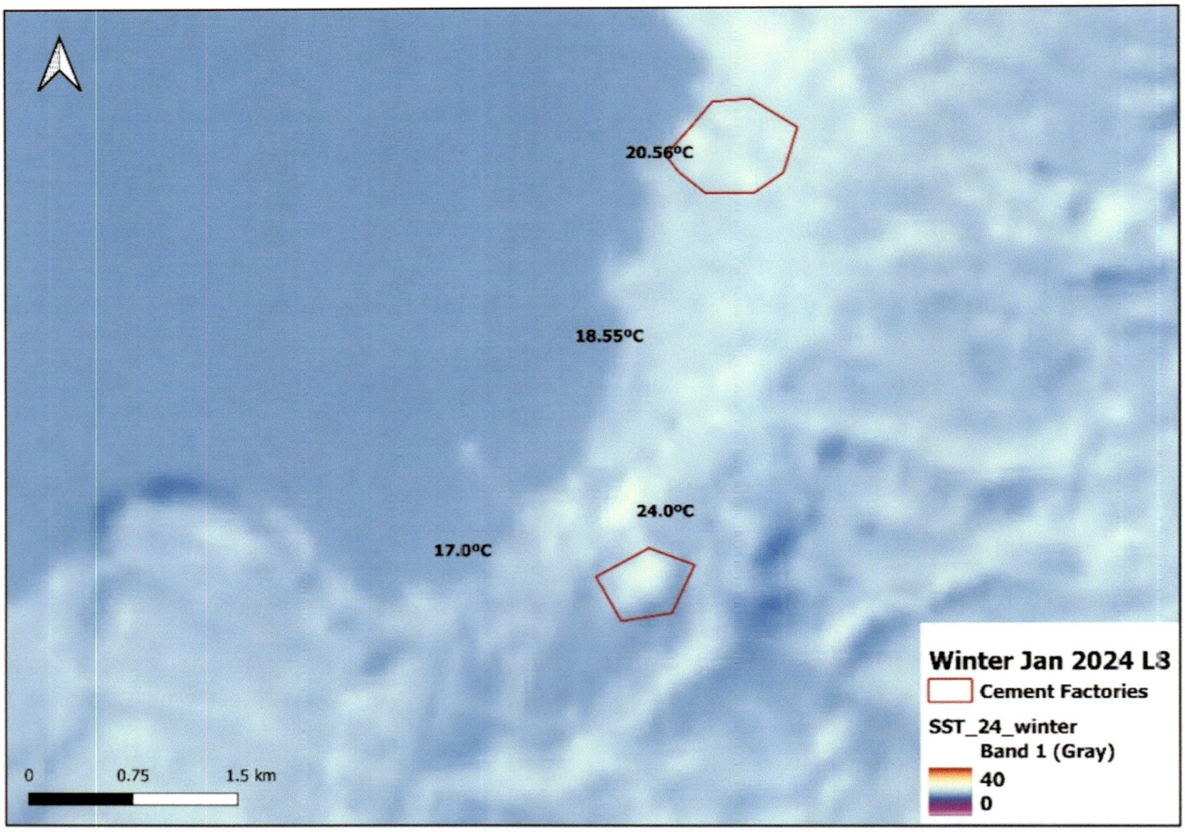

Figure 3.3: SST, January 2024. Copyright: Sirine Saad Eldine Ghiye.

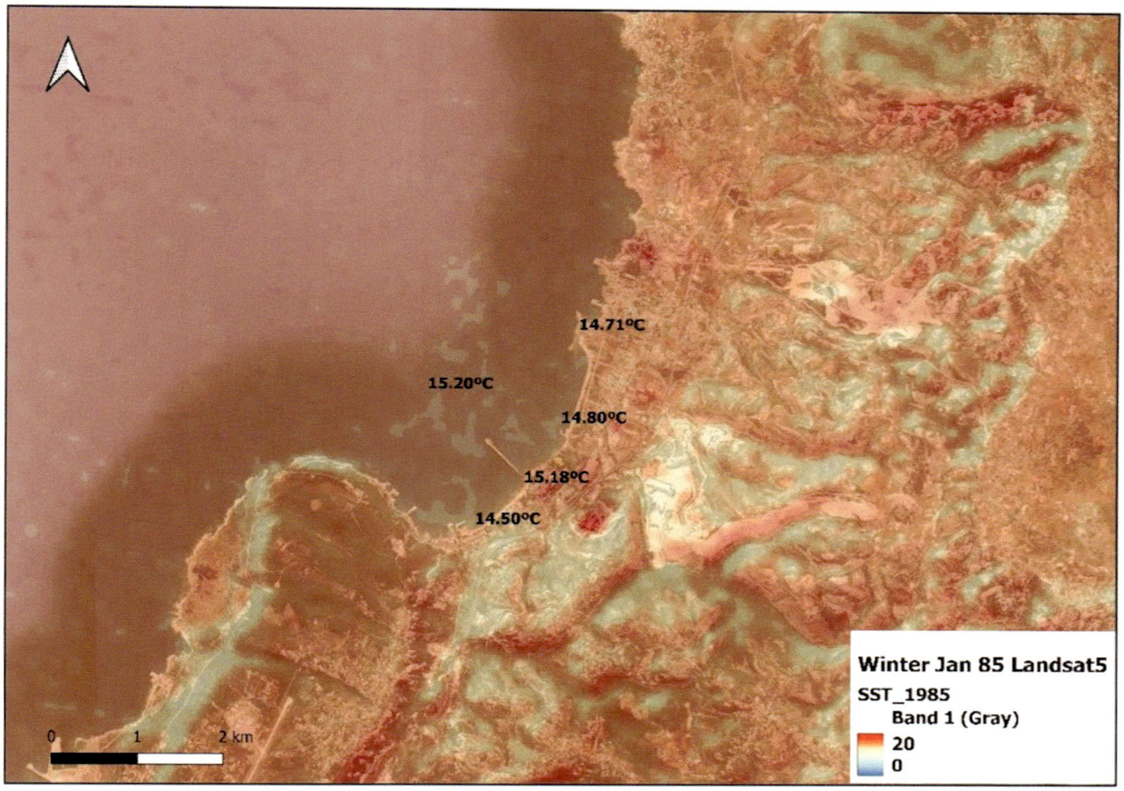

Figure 3.4: SST, January 1985. Copyright: Sirine Saad Eldine Ghiye.

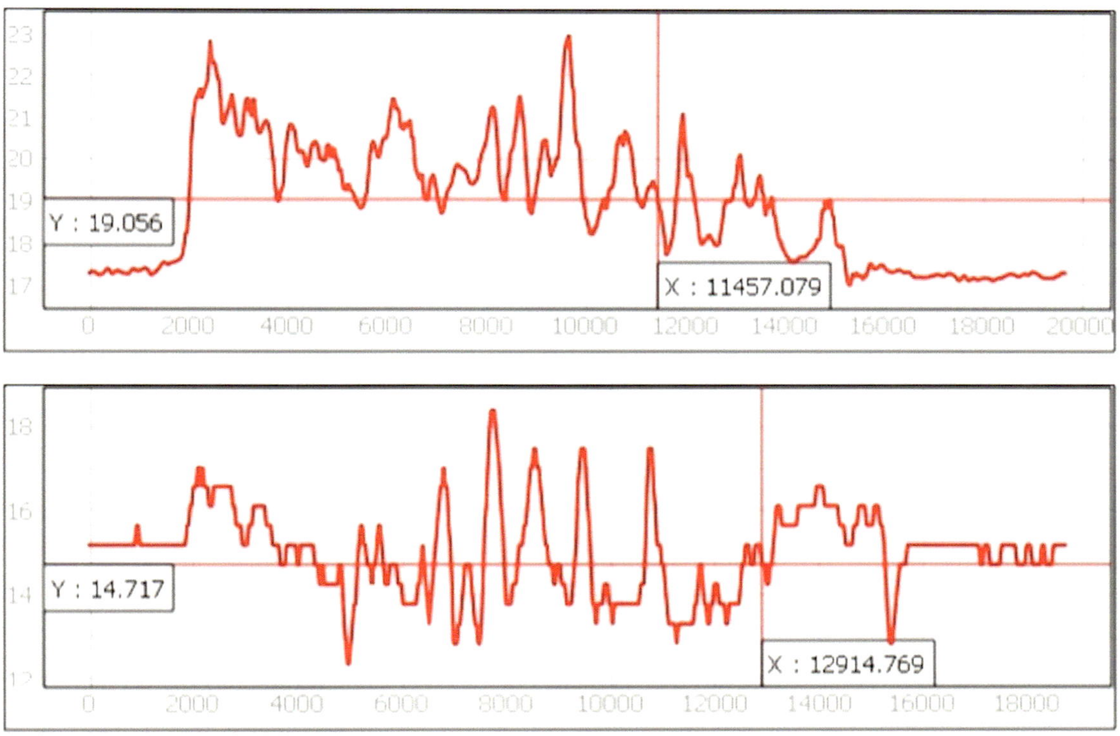

Figure 3.5: Temperature Profile, 2024 vs 1985. Copyright: Sirine Saad Eldine Ghiye.

Summer season

In Lebanon, the summer season typically extends from June to August, offering warm to hot weather with plenty of sunshine especially in coastal areas. Generally, the air temperature in those months ranges between 25°C to 35°C. The SST and the coastal land temperatures of Chekka were documented through satellite imagery for June of 1999 and June of 2023 (Figures 3.6-3.7). Analysis of the data indicates that in 1999 the temperature varied between 19 to 23°C, compared to 2023 when the temperature reached 28°C on the coast and 35°C near the cement factories. This change is also visible in the temperature profile presented in the graph of Figure 3.8, which shows that the maximum temperature was 26°C during 1999 compared to 36°C and more in 2023, while the average increases from 23.1°C to 34.5°C . It is probable that the temperature increase could be linked to emissions from cement factories, as well as loss of vegetation around Chekka, which occurred after 1999 (Ofreimu *et al.* 2025).

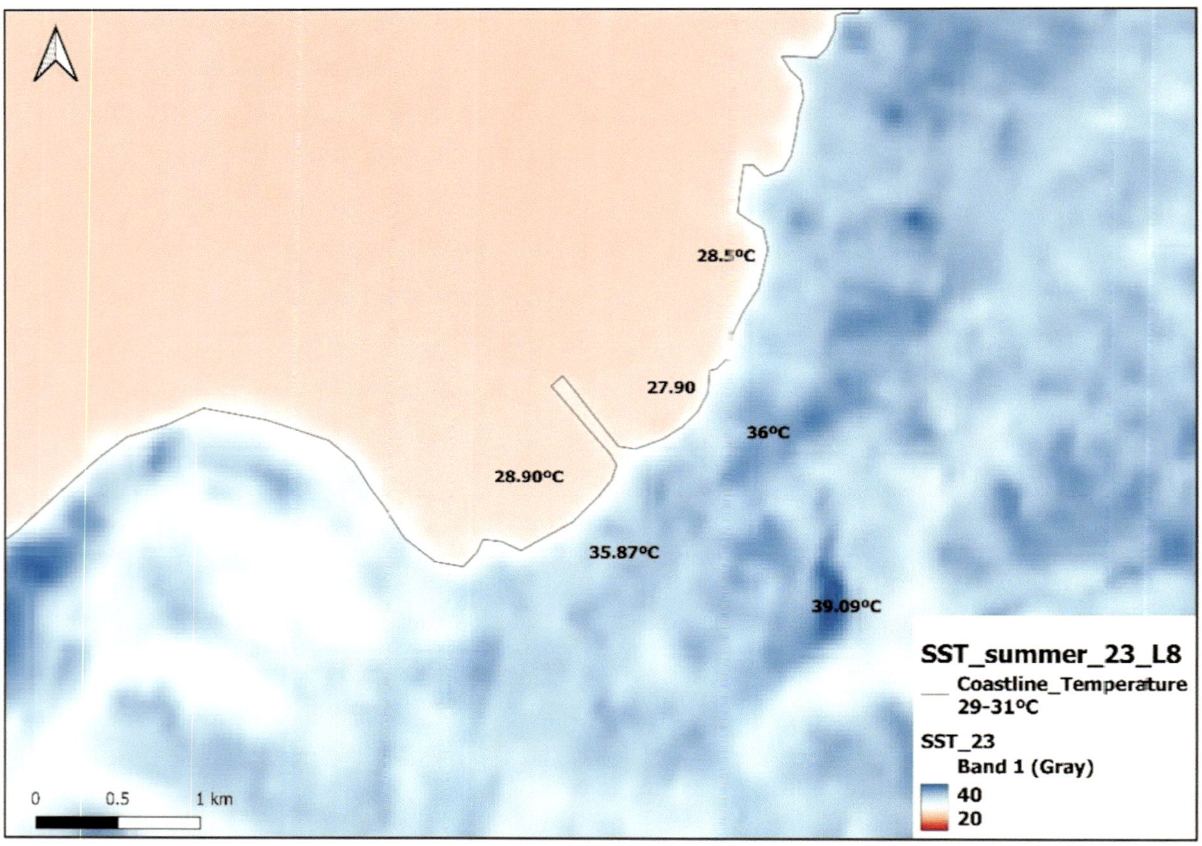

Figure 3.6: SST, June 2023. Copyright: Sirine Saad Eldine Ghiye.

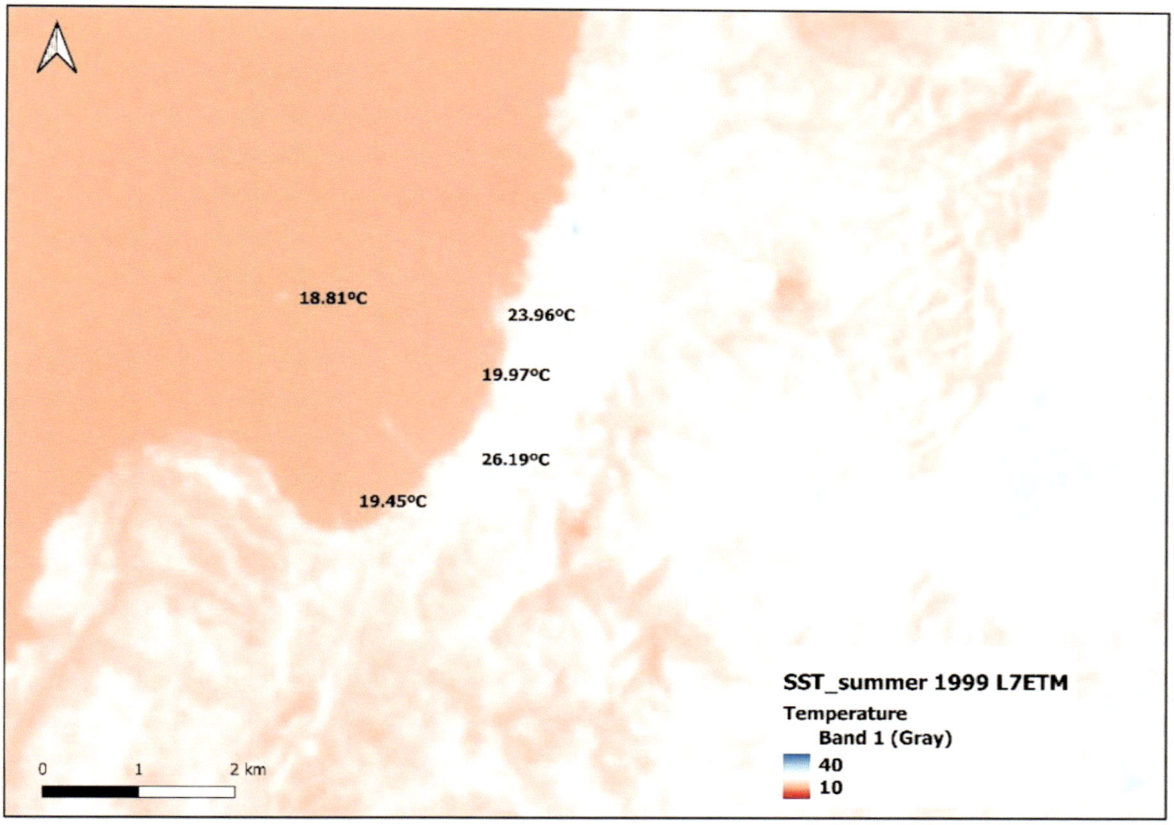

Figure 3.7: SST, July 1999. Copyright: Sirine Saad Eldine Ghiye.

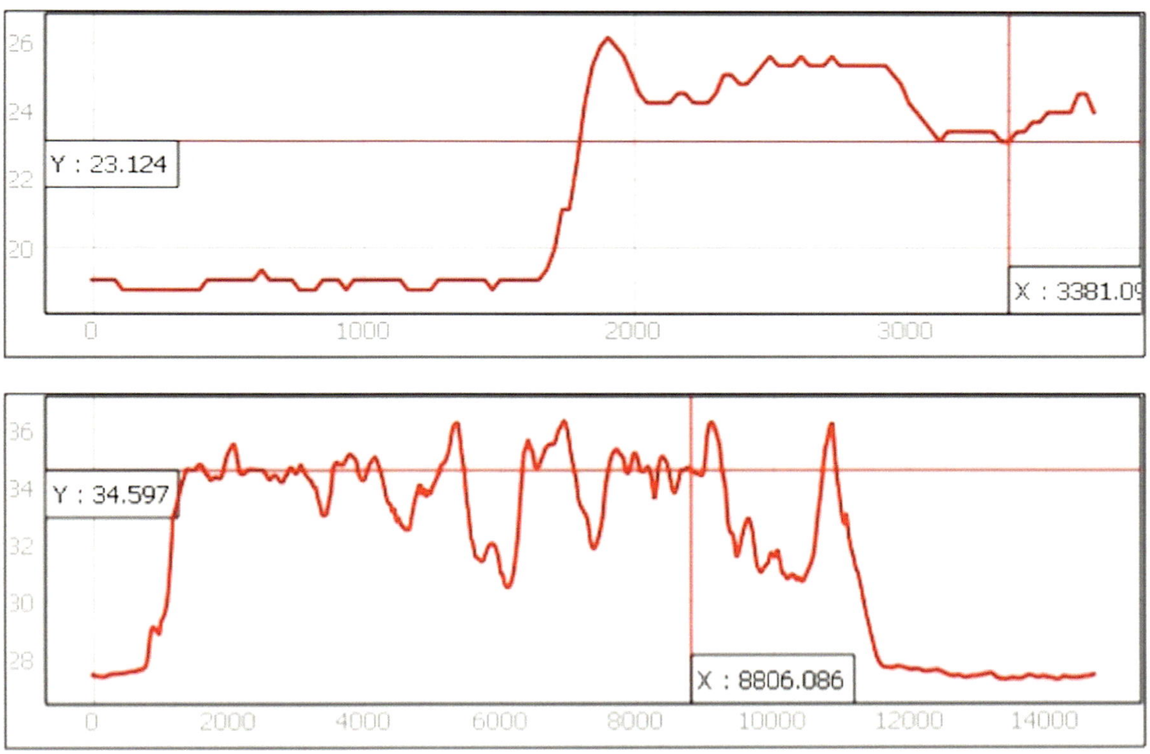

Figure 3.8: Temperature Profile, 1999 vs 2023. Copyright: Sirine Saad Eldine Ghiye.

Sea surface salinity in the last 20 years

Measuring salinity is essential for understanding various oceanographic processes and the distribution of marine life. Salinity also affects the site formation processes and the levels of preservation of underwater archaeological sites. Increased salinity leads to corrosion of metals associated with shipwrecks, and it also supports marine organisms which can damage submerged and coastal heritage assets. Recent studies have shown that sea surface salinity (SSS), measured either in parts per thousand (PPT) or practical salinity units (PSU), has increased in the eastern Mediterranean, including around Lebanon, due to climate change (e.g., Grodsky *et al.* 2019; Liu *et al.* 2025; Skliris *et al.* 2025).

Sea surface salinity was calculated as part of the study around Chekka. Figure 3.9 presents the salinity during 2022. Along the coastline the salinity ranged between 37-39 PSU, suggesting an increased risk of corrosion for underwater and coastal heritage with metal materials and changes to the marine environment. The increasing salinity affects the sediment of the seabed and therefore changes the marine biodiversity by supporting different algae and invertebrates, all of which can disturb submerged structures in the area.

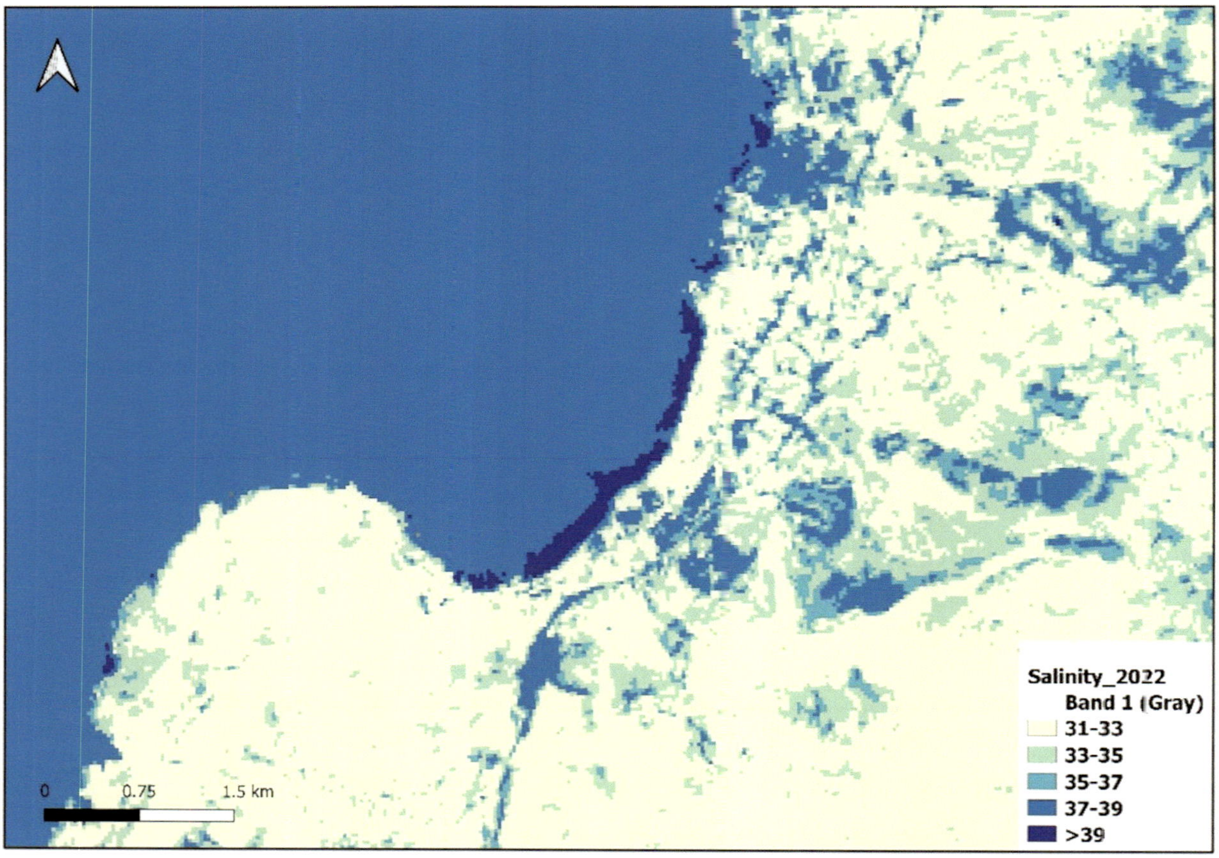

Figure 3.9: Quantity of salinity in Chekka and surrounding region. Copyright: Sirine Saad Eldine Ghiye.

Coastline changes

Analysis of satellite imagery in this study showed that there has been significant change to the coastline of Chekka during the last few decades, due to the processes of sea filling and accretion. At the same time, due to climate change the dynamic processes of the coast are increasing with a higher rate of erosion due to storms and unusual tides (Abou Dagher *et al.* 2012). In Chekka, there has been around

79,000m^2 of erosion of the sandy bank, which led to changes in the marine biodiversity and the geomorphology of the coastline (Figure 3.10). Erosion of the coastline in Chekka has also led to some loss of coastal features in the areas, where the ancient harbour must have been (El Safadi 2014), putting documented and undocumented archaeological remains at risk.

Figure 3.10: Coastline changes in Chekka between 1985 (purple line) and 2023 (red line) observed in satellite imagery. Copyright: Sirine Saad Eldine Ghiye.

Conclusion

As seen from the case study of Chekka, the use of satellite imagery and associated measurements of Sea Surface Temperature (SST) and Sea Surface Salinity (SSS), as well as observations and mapping of coastal changes over long periods of time, can provide valuable insights into the vulnerability of and risks to coastal and underwater cultural heritage assets. Currently, climate change has a significant impact on known and unknown maritime archaeological sites worldwide. In northern Lebanon, the increased sea temperatures, increased sea salinity and the coastal changes observed through the satellite imagery accessed for this project, show high levels of vulnerability for heritage assets in the seabed and on the coast, such as the not yet researched ancient harbour of Chekka. These findings underscore the growing concern but also the urgency to act in order to mitigate the threat that climate change poses to natural ecosystems, human societies and their heritage worldwide. This case study shows the relevance of maritime archaeology as a tool and research approach that can help track and understand changes in climate patterns and the environment, as well as human interactions and adaptations to the aquatic environments around them. Overall, it is hoped that the methodology proposed in this paper and the results from its application on the case study of Chekka will inspire further studies on the topic in Lebanon and abroad, while encouraging the consideration of sustainable development practices in maritime environments.

References

Abou Dagher M., M. Nader and S. El Indary 2012. Evolution of the coast of North Lebanon from 1962-2007: Mapping changes for the identification of hotspots and for future management intervention, IV International Symposium "Monitoring of Mediterranean Coastal Areas": Problems and Measurement Techniques At: Istituto di Biometeorologia (IBIMET), Consiglio Nazionale delle Ricerche (CNR), Italy.

Acharya, D. T. and I. Yang 2015. Exploring Landsat 8. International Journal of IT. *Engineering and Applied Sciences Research* 4: 4-10.

Advan U. and G. Jovanovska 2016. Algorithm for automated mapping of land surface temperature using Landsat 8 satellites data. Journal of Sensors: 1480307. <https://doi.org/10.1155/2016/1480307>

Ansari M. and M. Akhoondzadeh 2019, Mapping salinity using Landsat-8 OLI satellites images (Case study: Karun basin located in Iran). *Advances in Space Research* 65.5: 1490-1502. <https://doi.org/10.1016/j.asr.2019.12.007>

Bourouhou I. and F. Salmun 2021, Sea surface temperature estimation using remotely sensed imagery of Landsat 8 along the coastline of Tangier - Ksar Sghir region. *E3S Web of Conferences The International Conference on Innovation, Modern Applied Science & Environmental Studies* 234: 00096. <https://doi.org/10.1051/e3sconf/202123400096>

Boutin, J., N. Reul, J. Koehler, A. Martin, R. Catany, S. Guimbard, F. Rouffi, J. L. Vergely, M. Arias, M. Chakroun, G. Corato, V. Estella-Perez, A. Hasson, S. Josey, D. Khvorostyanov, N. Kolodziejczyk, J. Mignot, L. Olivier, G. Reverdin, D. Stammer, A. Supply, C. Thouvenin-Masson, A. Turiel, J. Vialard, P. Cipollini, C. Donlon, R. Sabia and S. Mecklenburg 2021. Satellite-based sea surface salinity designed for ocean and climate studies. *Journal of Geophysical Research: Oceans* 126.11: e2021JC017676. <https://doi.org/10.1029/2021JC017676>

El Safadi C. 2014. Bronze Age and Iron Age Levantine harbours: an Evaluation of their Maritime Accessibility and Protection. Unpublished Master's dissertation, University of Southampton.

Ghiye, S. S. E. 2023. Potential Climate Change Threats to underwater cultural heritage, in J. Craig J. Mok, T. K. Knoerl, E. Perez-Alvaro and O. Varmer (eds) *Proceedings of Fifth Asia Pacific Regional Conference on Underwater Cultural Heritage: Gwangju, The Republic of Korea, November 13-18 2023. The Museum of Underwater Archaeology*. <https://apconf.omeka.net/exhibits/show/apconfproceedings/item/1960>

Gregory, D., T. Dawson, D. Elkin, H. Van Tilburg, C. Underwood, V. Richards, A. Viduka, K. Westley, J. Wright and J. Hollesen 2022. Of Time and Tide: The Complex Impacts of Climate Change on Coastal and Underwater Cultural Heritage. *Antiquity* 96.390: 1396–1411. <https://doi.org/10.15184/aqy.2022.115>

Grodsky S., N. Reul, A. Bentamy, D. Vandemark and S. Guimbard 2019. Eastern Mediterranean salinification observed in satellite salinity from SMAP mission. *Journal of Marine Systems* 198: 103190. <https://doi.org/10.1016/j.jmarsys.2019.103190>

Global climate observation system, viewed 10 October 2025, <https://gcos.wmo.int/site/global-climate-observing-system-gcos/essential-climate-variables>.

Kennedy V., R. R. Twilley, J. A. Kleypas, J. H. Cowan Jr and S. R. Hare 2002. *Coastal and marine ecosystems and global climate change - Potential Effects on US Resources.* Washington DC: Pew Research Centre.

Kim, Y. J., D. Han, E. Jang, J. Im and T. Sung 2023. Remote sensing of sea surface salinity: challenges and research directions. *GIScience & Remote Sensing* 60.1. <https://doi.org/10.1080/15481603.2023.2166377>

Kokal A. T., I. Ismailoglu, N. Musaoglu and A. Tanik 2022. Detection of surface temperature anomalies of the Sea of Marmara. *Advances in Space Research* 71.7: 2996-3004. <https://doi.org/10.1016/j.asr.2022.10.055>

Lange M. A., 2020, Climate Change in the Mediterranean: Environmental Impacts and Extreme Events, European Institute of the Mediterranean (IEMED), viewed 10 October 2025, <https://www.iemed.org/publication/climate-change-in-the-mediterranean-environmental-impacts-and-extreme-events/>.

Liu C., X. Liang and L. Yu 2025. Salinity Trends and Mass Balances in the Mediterranean Sea: Revisit the Role of Air-Sea Freshwater Fluxes and Oceanic Exchange. *Ocean Science* 21.5: 2069–2083. <https://doi.org/10.5194/os-21-2069-2025>

Marghani M., and M. Hashim 2011. A numerical method for retrieving sea surface salinity from MODIS satellite data. International Journal of the Physical Sciences 6.13: 3116-3125. p. 3116-3125. <http://www.academicjournals.org/IJPS>

Ofremu G., B. Y. Raimi, S. O. Yusuf, B. A. Dziwornu, S. G. Nnabuife, A. M. Eze and C. A. Nnajiofor 2025. Exploring the relationship between climate change, air pollutants and human health: Impacts, adaptation, and mitigation strategies. *Green Energy and Resources* 3.2: 100074. <https://doi.org/10.1016/j.gerr.2024.100074>

Reul N., N. Reul, S.A. Grodsky, M. Arias, J. Boutin, R. Catany, B. Chapron, F. D'Amico, E. Dinnat, C. Donlon, A. Fore, S. Fournier, S. Guimbard, A. Hasson, N. Kolodziejczyk, G. Lagerloef, T. Lee, D.M. Le Vine, E. Lindstrom, C. Maes, S. Mecklenburg, T. Meissner, E. Olmedo, R. Sabia, J. Tenerelli, C. Thouvenin-Masson, A. Turiel, J.L. Vergely, N. Vinogradova, F. Wentz and S. Yueh 2020. Sea surface salinity estimates from spaceborne L-band radiometers: An overview of the first decade of observation (2010–2019). *Remote Sensing of Environment* 242: 111769. <https://doi.org/10.1016/j.rse.2020.111769>

Skliris N., R. Marsh, M. Breedon and S.A. Josey 2025. Accelerated Warming and Salinification of the Mediterranean Sea: Implications for Dense Water Formation. *Journal of Marine Science and Engineering* 13.1:25. <https://doi.org/10.3390/jmse13010025>

Spalding M., 2011. Perverse Sea change: underwater culture heritage in the ocean is facing chemical and physical changes. *Cultural Heritage & Arts Review.* <https://oceanfdn.org/sites/default/files/Perverse%20Sea%20Change%20MJS1.pdf>

US Geological Survey, Landsat mission, viewed 14 January 2024, <https://www.usgs.gov/landsat-missions>.

Yang J., P. Gong, F. Rong, M. Zhang, J. Chen, S. Liang, B. Xu, J. Shi and R. Dickinson 2013. The role of satellite remote sensing in climate change studies. *Nature Climate Change* 3: 1001. <https://doi.org/10.1038/nclimate2033>

Zhou, D., J. Xiao, S. Bonafoni, C. Berger, K. Deilami, Y. Zhou, S. Frolking, R. Yao, Z. Qiao, Z. and J.A. Sobrino 2019. Satellite Remote Sensing of Surface Urban Heat Islands: Progress, Challenges, and Perspectives. *Remote Sensing* 11.1: 48. <https://doi.org/10.3390/rs11010048>

4.
Contemplating the Future of a Complex Coastal Heritage Asset in the Face of Climate Change: The Case Study of Whitford Point Lighthouse

Bella Romain

This project examines conservation options for coastal heritage assets at risk from climate change effects through the use of Multi-Attribute Value Theory (MAVT). MAVT is a Multi-Criteria Decision Analysis (MCDA) technique which supports decision making in scenarios where the preferred intervention is determined by its overall performance against several ranked criteria. Even though MCDA has previously supported decision making for redundant heritage sites earmarked for reuse, MAVT has not been applied to conservation management strategies for heritage at risk. MAVT's capacity to objectively evaluate qualitative and quantitative values was the basis for its selection for this research. The process provides a holistic framework for the assessment of conservation interventions for heritage at risk of loss. MAVT refers to conservation interventions as 'alternatives' and the criteria as 'attributes'.

In this paper, MAVT has been adapted and applied to Whitford Point Lighthouse in Gower, Swansea, Wales, UK, to make a hypothetical decision about its future. This process has been limited by restricting the weighting of criteria only to experts and not including affected communities. The interventions (including maintenance, minimal intervention, and restoration) were evaluated by the MAVT process, and the results indicated that adaptive release (also described as 'planned release') is the most sustainable approach. Restoration was a strong contender in sensitivity tests but is unfeasible due to the prohibitive costs.

Introduction

This research aims to develop and assess an objective method to support decision making for heritage at risk. Specifically, in this project, Multi-Attribute Value Theory (MAVT) is adapted for use as a potential tool to navigate the highly controversial and emotive issue of the futures of heritage at risk. Heritage at risk refers to any heritage asset, (e.g., a building, monument, or archaeological remains), which faces a substantial risk of loss or dereliction for any reason. Factors which compound a heritage asset's 'at-risk' status include lack of repairs and maintenance, the cost of these ongoing, limited revenue potential, and, increasingly, anthropogenic climate change. In the UK, registers of heritage at risk are compiled by both amenity societies (e.g., SAVE Britain's Heritage, The Georgian Group, The Victorian Society), who compile data to engage the public, monitor progression, and put pressure on owners, and statutory bodies (e.g., Historic England, Historic Environment Scotland), who have a duty to identify, assess, and monitor vulnerable heritage. Within this paper, the focus is on risks which are both significant and specific to Whitford Point Lighthouse in South Wales. Most of these risks and possibilities discussed, although contextually different, can be applied to many other heritage assets.

In Wales, monitoring buildings at risk is overseen by Cadw, who commissions contractors to assess the overall condition of all listed buildings and scheduled monuments over a five-year period. Cadw provides a register of local at-risk buildings to local authorities, who have ultimate responsibility for ensuring ongoing protection. Although there is currently no publication of a national buildings at risk register, some councils publish local registers, and a recent Freedom of Information request resulted in the release of point-in-time Welsh Buildings at Risk data (Welsh Government 2023; 2024).

Some conventional fabric-based options for heritage at risk include restoration, maintenance, and minimal intervention (see Table 4.1). The feasibility of each of these is influenced by economic viability, including the possible need to overcome conservation deficit (where the costs involved in repairing an asset far exceed its income generating potential), current condition, and other factors, including environmental implications and accessibility. Although some assets may be readily saved, others present challenges which either demand considerable compromises to be made or relegate the asset to dereliction. Recently, in the face of anthropogenic climate change, critical heritage proponents have begun to explore more intangible solutions to heritage at risk, recognising the value that other modes of recording and interaction contribute to the heritage of a site, beyond the asset itself.

Table 4.1: Impact of interventions to heritage assets, adapted from De Silvey *et al.* 2022. Copyright: Bella Romain.

Intervention	Impact	Description	Outcomes
Maintenance		Retain asset in appropriate use.	Maintenance, timely repair, sympathetic alterations to prolong use.
Conservation		Significant heritage value which warrants investment.	Maintenance, repairs and appropriate restoration using traditional techniques and materials. Maximum retention of original fabric and minimal intervention.
Adaptive reuse		Changes allowed to enable continued use.	Alterations or conversion dictated by significance.
External protection		Structure protected to prevent further decay. No access possible.	Mothballed until funds or techniques to conserve are available.
Relocation		Structure moved to enable ongoing conservation.	Harm to context and to structure likely. Options will be limited by significance.
Managed loss/ decline		Asset is no longer significant enough to warrant conservation, or logistics of conservation prohibit action.	No maintenance, leading to eventual ruin. May be stabilised or demolished in future.
Adaptive release		Transformation through management of decay. Recording and monitoring change, embracing opportunities for historical, environmental, or social benefit.	Preservation by record followed by engagement with the change, determined by resources and community engagement. Regular monitoring and review.

Current policy documents and adaptation plans (e.g., Fluck 2016; National Trust 2007) recognise and reiterate the need to find alternative and sustainable conservation solutions to the inevitable loss of heritage that will occur due to anthropogenic climate change. However, despite these calls to modify approaches to conservation from the physical to the intangible, and to incorporate adaptation and loss frameworks (Venture *et al.* 2021: 406), there are no recommendations on how this should be managed, nor suggestions for identifying those assets which may be tolerably lost. Planned loss is not a unilateral solution to heritage assets which are at risk. There are many which can be economically reused, adapted, or sensitively restored. It is important, however, to accept and understand that, in a small handful of cases, loss is the only viable option.

UK conservation policy is, for the most part, still focused on the preservation of built fabric, and its loss is seen as the ultimate failure, rather than an opportunity for archaeology or understanding (Avrami 2020: 229; Khalaf 2022: 343). From this viewpoint, intangible heritage is deemed to have less value than the physical asset, no matter its condition, and intangible values that a site has may go unobserved and unrecorded (Emerick 2014: 4-5). If we can move away from this perspective, towards one which values diverse types of heritage, we might come to see how loss is not an end point, but both transition and legacy.

The need for a framework to move adaptation from a policy and practice buzzword to a real-world approach to critical heritage at risk is undeniable, not least because total loss, rather than physical loss, will be the reality without guidance and the confidence to apply it. With this need ascertained and accepted, the question then becomes one of how 'losable' heritage assets are identified, how loss might be quantifiable, what heritage can be created in the losing, and how this process can be transparent, justifiable, and fair.

Multi-Criteria Decision Analysis (MCDA) is a means of making complex decisions with multiple criteria. Specifically, Multi-Attribute Value Theory (MAVT) is a MCDA method in which relevant experts assign weights to attributes to indicate their importance in the decision-making process. Each alternative is scored for its performance against each of the attributes. The alternative with a score closest to 1 is the preferred alternative when all of the attributes are considered. As a compensatory process (i.e., a high score in one area can compensate for a low score in another), the tool can be tailored to prioritise specific attributes (e.g., cost, environmental impacts, conservation benefit) through weighting. MCDA has been used reliably in supporting decision making for built heritage planning and heritage management, for example to identify a particular building for a specific re-use potential, or to identify which building would be best suited for one of several re-uses (Ferretti *et al.* 2014), but the author did not find any examples of MAVT being used to calculate the preferred intervention for a heritage asset at risk.

The use of MCDA techniques is recommended for supporting, rather than making, decisions. Whilst the process, if fairly applied, should deliver a balanced outcome, there is opportunity to skew the results in favour of specific objectives. As such, the results of the process should be viewed as a comparative assessment of the options rather than a definitive decision in itself.

To contextualise and determine the use of MAVT as a decision-making support tool, this study will critically examine and assess common conservation interventions against the specific circumstances of Whitford Point Lighthouse. Relevant risks to the lighthouse, the considerations and limitations of its setting, and the professional opinions of both expert stakeholders and statutory bodies contribute to this process.

Whitford Point Lighthouse is an internationally significant Victorian cast-iron wave-washed tower lighthouse; one of only two built globally. Completed in 1896 and decommissioned in 1920, the subsequent lack of maintenance has led to a state of perilous dereliction. Both access difficulties and current environmental protections make conservation challenging and expensive. The lighthouse is unusual in that it has dual-designation as both a scheduled monument and a listed building, a factor which complicates the application of conservation legislation. Whitford Point is dual-designated due to its rarity, its significance as a structure, and its role in industry and maritime engineering. Over a century of neglect has ensured that any interventions will be costly, logistically and philosophically challenging, and of uncertain longevity. These complexities, coupled with the uncertainty of its future, made Whitford Point Lighthouse a strong candidate for a real-life study of the MAVT process.

To consider the effectiveness of this MAVT tool in decision making, the paper starts by exploring the range of conservation interventions available to built heritage, and the attributes of a site which have an impact. Subsequently, the paper focuses on the significance of heritage assets and how this influences the conservation choices made by heritage managers, alternative conservation interventions available for heritage assets, the impact of these and the characteristics of sites and their environments which may influence the possibility of conservation interventions. Following that, the methodology of Multi-Attribute Value Theory (MAVT) and the prioritisation screening assessment for this project is explained. Finally, the case study of Whitford Point Lighthouse is presented, including the application of MAVT in determining a best-case scenario for its future.

Conservation Management of Heritage at Risk

Determining conservation options

Conservation is often a controversial subject between those who have professional and academic expertise within the area. This conflict of interest increases when it comes to differences of opinion between those with no theoretical understanding of conservation principles and those, both professionals and members of the community, who do (Walker 2011: 25). In more detail, an asset in use is more likely to survive, especially if it can generate enough income to be sustainable. Routine use of a building also kindles greater community value, in turn engendering public interest in its continued survival (Avrami 2009: 179). Such a building is less likely to be subject to acts of vandalism and become an eyesore. Assets unable to maintain their original function are often at risk due to vacancy and lack of maintenance. While reuse of vacant buildings is becoming a popular climate mitigation proposal, with the additional benefit of repurposing historic buildings, there are specific requirements and compromises which must be made to ensure a building's future. Timely maintenance is essential in preventing decay (Cadw 2011: 22; Casey and Becker 2019: 176) and although there is technically no legal requirement to maintain listed buildings (Historic England 2013), there is an expectation that condition will be sustained or enhanced. Despite being better protected in other ways, scheduled monuments do not enjoy the same expectation, as change and decay are inevitable, and their conservation is the management of decline (Emerick 2014: 17). For assets with dual-designation, scheduled monument status takes precedence (Planning (Listed Building and Conservation Areas) Act 1990 S61). As with scheduled monuments, for many 'at-risk' assets, conservation is concerned with the management of decline. Effective conservation of heritage at risk is discussed throughout conservation literature (e.g., Dawson 2016; De Blieu and Halminski 2001) but is the focus only in rare cases (Dawson *et al.* 2021; Guest *et al.* 2022), usually in considering the effects of anthropogenic climate change on coastal heritage assets (Casey and Becker 2019; Hassan and Xie 2020).

Wienberg (2014: 70) suggests that buildings acquire heritage value when faced with a threat. These assets are given a 'thrive or die' ultimatum, and opportunities to make the most of the decay process to understand a structure and its history, which would not be permissible in normal circumstances, are not taken. Weinberg (2014: 74) explores a theory of 'deconstructive archaeology' to shift the perception of heritage asset loss from negative to positive, in a similar vein as Dawson *et al.* (2021) and DeSilvey *et al.* (2021), who use recording to increase community engagement with heritage. Industrial and maritime heritage assets are more likely to fall into an 'at-risk' category than other assets and are less easily repurposed due to their uses or locations (Colliers International 2011: 23). For the many softer industrial buildings (mills, pump houses, land-based lighthouses) which have been converted to residential, administrative and hospitality uses, there are more standing ruins which have no viable future.

Whilst all heritage could be considered 'at-risk', when a building is placed on a register, it implies that actual harm is happening, and the risk is of eventual or imminent loss, rather than degrees of hypothetical harm. The nature and probability of the risk is a principal factor in why the building is vulnerable, as well as if, and how, it might be saved. Some risks are common to all heritage assets, including a lack of maintenance (ICOMOS 2020), the cost of repairs or lack of funding (Nevell 2010), institutional or competency barriers (Fatorić and Seekamp 2017), and the impact of policies and legislation (i.e., administrative burden) on moving from plans and consent to completed works. Additionally, the timing of grant funding rounds often requires money to be spent out of season and at an unrealistic pace to secure contractors. Assets also face context specific risks including inappropriate or encroaching developments, environmental pollution, conflict, and anthropogenic climate change. Whilst some harm (e.g., flooding) can be mitigated against, others (e.g., increased storminess) are much harder to predict or control. Some risks are in the form of site characteristics, such as location, accessibility, or environmental protections, which limit the type of access, or works, that can be conducted *in situ*. Additionally, if access is an issue, funding might not be available for works due to limited public benefit. All of these must be considered when selecting appropriate conservation interventions as they result in higher costs and limitations, which affect the success of any planned project.

Significance

Conservation philosophy is based on decades of evolving relationships between people, places, and things. Charters have developed to represent increasing awareness of heritage types, and the values people ascribe to them (Emerick 2014). These form the basis of legislation and guidance, and the subsequent management of heritage assets. Welsh guidance (Cadw 2011) recognises the importance of a landscape-based approach to conservation and the need to understand the value of a site's historic and natural significance (Cadw 2011: 6), recommending that this should underpin decisions.

Significance is defined by a series of values:

Historic	Associations with people or events of note.
Evidential	Evidence of past materials, methods, technologies.
Aesthetic	The intended design or resultant patina; how it looks.
Communal	What it means to people.

The impacts, both positive and negative, of potential interventions on the values and significance of a site, need to be assessed as part of this process (Cadw 2011: 9).

Interventions

For most heritage assets, intervention is in the form of ongoing maintenance and occasional repair. As assets become vulnerable, interventions tend to have a higher impact and must be justifiable as determined by the risk of loss as an alternative. Some broad definitions of common interventions are given below.

Approaches to the conservation of heritage assets are included in conservation management plans, which focus on preservation of fabric (e.g., Kerr 2013) using International Council on Monuments and Sites' (ICOMOS) charters as the foundation. Forster and Kayan (2009: 214) argue that whilst heritage at risk registers in the UK have low efficacy due to the multiple curators, systems and means of assessment, management plans help to avoid buildings becoming 'at-risk' by scheduling maintenance and reducing the expense (and impact on significance) associated with *ad hoc* repairs. Emerick (2014: 195) suggests that the process of creating a conservation management plan is as valuable as the final document. The activity of deciding what to save and why is integral and rooted in the significance and values of assets (Avrami 2009; Wienberg 2014: 74).

The aim of conservation is to retain or recover the cultural significance of a site (Matero 1993: 17) and any measures must critically consider how they impact those values (Cadw 2011: 9). This paper proposes an adapted MAVT tool to support decision making, balancing the diverse and complex requirements of the site with the conservation requirements demanded by such a unique asset.

Minimal Intervention

Minimal intervention is often eulogised as the 'epitome of conservation practice' and is the foundation of the SPAB (Society for the Protection of Ancient Buildings) approach where it is partnered with a system of regular maintenance (Morris and Webb 1877). True minimal intervention is allowing a building to decay unchecked (Muños Viñas 2009: 49). Used as a preventative, it should be timely, using proven techniques and avoiding unnecessary replacement of materials (Matero 1993: 18).

The historical approach to preserving scheduled monuments was to 'freeze in time'. Later, building recording and adaptive reuse became the preferred option (Chitty 1987: 52). Today the usual methodology for scheduled monuments is 'managed decline' rather than prevention (DeSilvey *et al.* 2021: 420), aiming to 'slow down...dilapidation through limited maintenance' (Storm 2008: 33), and minimising impact on historic fabric and character (Forster and Kayan 2009: 212). Repairs, as opposed to maintenance, have the potential to cause damage to historic material and the authenticity of the structure (Forster and Kayan 2009: 211; Muños Viñas 2009: 49).

Restoration

Restoration is defined as 'returning a place to a known earlier state by removing accretions or by reassembling existing elements without the introduction of new material' (The Burra Charter, ICOMOS 2013: 2). Cadw (2011: 26) stipulates conditions, based on guidance from conservation charters, in which it considers restoration to be acceptable. Restoration should only happen when other paths of conservation are impossible, loss is the alternative, and evidence is available to inform decisions. It should not prejudice the possibility of future learning or interventions for the heritage asset. Despite the necessity of its placement in the arsenal of conservation options, restoration remains controversial (Stanley-Price 2009: 32) as reassembly of existing elements can often only occur with the introduction of new material (Emerick 2014: 186). Reconstruction is considered acceptable only in exceptional cases, for example when the alternative is total loss (Riga Charter, ICCROM 2000: A6; Stanley-Price 2009: 182).

The greatest challenge for restoration is the preservation of 'authenticity' (Earl 2003: 70), a term often used but inadequately defined for its practical application. The Athens Charter (ICOMOS 1931: A2) suggests that authenticity is conferred by 'material that is original and genuine', while the Nara Document (ICOMOS 1994: A13) states that authenticity is determined by the cultural values of the place. The Burra Charter (ICOMOS 2013: 3) prioritises 'spirit of place' over fabric but maintains a stance which privileges material. Cadw's (2011: 9) 'Conservation Principles' states that the authenticity and future significance of a heritage asset should be retained but does not define 'authenticity'. Authenticity is considered to be synonymous with age and patina (DeSilvey and Harrison 2019: 2; Matero 1993: 17) and maintaining legibility of 'design, materials, workmanship and setting' (Emerick 2014: 174). Integrity tends to be more commonly used outside of Europe and is somewhat easier to define as the link between people, story, and place (Emerick 2014: 216). It is generally recognised that achieving authenticity of fabric is impossible (Stanley-Price 2009: 37) and that even minimal intervention approaches result in an overall loss of authenticity over time (Chitty 1987: 52). This loss, implied or otherwise, of authenticity and original fabric is perpetually seen as negative and results in practitioners' inability to objectively consider restoration and reconstruction as viable options (Avrami 2020: 216). The human desire for continuity of presence, familiarity, and a sense of the known, has little to do with the concept of authenticity as defined by professionals (Emerick 2014: 178; Jackier 2001: 92; Piazzoni 2020: 241).

Translocation

Whilst the practice of moving buildings has died out in the UK, it remains popular in the USA and Australia due to the continued practice of building with lightweight timber frames (Drozd 2019: 177) and the lower costs of transportation compared to new buildings (Jackier 2001: 45). Translocation provokes strong opinions. Reasons against the movement of heritage assets include the elevated risk of loss of fabric and authenticity, logistical challenges, and expense (Hassan and Xie 2020: 23). Others believe current policy portrays relocation as objectionable (Peltola 2008: 1) and practitioners are ill-equipped to critically evaluate alternatives to traditional preservation (Casey and Becker 2019).

Both historic landscape characterisation and marine character area assessments play a role in understanding the relationships between heritage assets and their settings (Cook 2019: 179). It is therefore important to critically examine whether the potential risk to this aspect of significance is worth definite loss of the entire asset (Peltola 2008: 1) or whether its physical survival is worth damaging all other values (Emerick 2014: 235). The impact of a move is borne both by the asset and the landscape, where the removal of an asset leaves a hole (Goblet 2006: 35) or conversely, its introduction changes the existing *genius loci* elsewhere (Heesom *et al.* 2021: 2953). Transportation of large structures intact presents greater challenges and risks, but some recommend it as the preferred method due to maintenance of integrity (Peltola 2008: 123). Moving following partial or total disassembly allows for repairs to be made but original fabric and embodied craftsmanship is lost (Heesom *et al.* 2021: 2948). Relocating buildings is seen as objectionable unless the alternative is loss (Gregory 2008: 114; Peltola 2008: 1). This implies that tangible heritage has greater significance than anything that might remain, and the risk of inevitable loss should prompt relocation (Historic Environment Group 2020: 14; Jackier 2001: 88), particularly when the heritage asset in question is rare (DeBlieu and Halminski 2001: 177; Drozd 2019: 177).

Translocation can be seen as an opportunity for 'destructive archaeology' (Weinberg 2014: 71), followed by reconstruction, linking past and present (Khakpan *et al.* 2020: 54). During disassembly, Historic Building Information Modelling (HBIM) can be utilised to scan and place elements (Heesom *et al.* 2021: 2960). This allows for accurate rebuilding of the structure and ensures that, should any losses occur, there is a full record that can be used to recreate them. HBIM provides ways to safeguard authenticity without necessarily preserving physical fabric. Point-cloud recording can generate 3D models allowing virtual walkthroughs and enhancing accessibility, especially for heritage assets in remote locations, such as the Whitford Point Lighthouse, the case study of this project, or for digitally restoring heritage assets, which can either be 3D-printed or used as interactive interpretation. It is generally advised, that in translocation heritage assets should be relocated within their setting, or to a comparable location, ideally with an appropriate reuse (The Burra Charter, ICOMOS 2013: 4), or, alternatively, moved to open air museums.

Adaptive Release

Adaptive release (also described as 'planned release') is another conservation approach to change and loss of heritage assets (DeSilvey and Harrison 2019: 2). Adaptive release is a landscape-scale intention to work with, rather than against, dynamic transformations (DeSilvey *et al.* 2021: 418). If conservation is management of change, then, by definition, it must include the management of loss. Loss confers meaning and value, often to things which had no apparent value before their loss was anticipated (Wienberg 2014: 70). As a society, perceptions of loss are negative, but adaptive release has been considered by some as 'generative...facilitating the emergence of new values...and forms of significance' (DeSilvey *et al.* 2021: 421; Khalaf 2022: 351).

Adaptive release is particularly relevant to coastal heritage environments, where assets are at risk of loss due to climatic changes. Although policy documents acknowledge the need to evaluate current preservationist approaches and to accept and plan for loss in the face of climate change (Fluck 2016: 21; Historic Environment Group 2020: 9), frequently 'preoccupation with...fabric' prevents relevant organisations from doing so (Khalaf 2022: 343).

Consideration of landscape-scale loss and the integration of divergent priorities demand a transdisciplinary approach with opportunities to contribute to the process (Gregory *et al.* 2022: 1407; Historic Environment Group 2020: 16; Venture *et al.* 2021: 400). Policies should provide a scaffold to navigate the difficult 'dichotomy between best practice and reality, between hypothetical plans and what we can bear to lose' (Gregory 2008: 112). Planning for loss is difficult because it is unpredictable and rates of decay vary (DeSilvey *et al.* 2021: 418). The aim of adaptive release is to proactively anticipate the loss of heritage assets as the starting point for engaging with communities to prepare for, and observe, the process of losing that asset. This reversal of perspective transforms loss from a failure to conserve, into an opportunity to create heritage, tangible and intangible (Avrami 2020: 215; DeSilvey 2006: 324; DeSilvey *et al.* 2022: 5; Khalaf 2022: 349; National Trust 2007: 15; Venture *et al.* 2021: 406).

Recording and monitoring the asset and active stakeholder engagement can be used to develop a management plan for heritage at risk where traditional preservation techniques are judged unsustainable (Guest *et al.* 2022: 2-3) or when there are no viable alternatives (DeSilvey *et al.* 2022: 24). Encouraging communities to take custody of their heritage through archaeological activities and monitoring can facilitate conversations about loss and foster a sense of ownership (Dawson *et al.* 2021). For example, accessibility has been increased by the development of apps for recording and monitoring sites (Smith 2014). Moreover, encouraging the observer to imagine past histories (Cooke 2000: 146), which Emerick (2014: 150) argues, maintains heritage as much as lost fabric, and evokes creative responses from participants (Emerick 2014: 9). Projects such as Scotland's 'SCHARP' and 'Shorewatch' engage communities with heritage at risk through citizen science (DeSilvey *et al.* 2022: 4; Historic Environment Group 2020: 18) and alleviate pressures on archaeology services (Dawson *et al.* 2021). Similar coastal monitoring projects exist in Wales (Pembrokeshire Coast National Park 2023; Welsh Coastal Monitoring Centre 2023). Despite its significant contribution to dealing with loss, adaptive release is not suitable for heritage assets with a viable future and adequate funding (DeSilvey 2023). Hence, an asset's significance, expected service-life and the sustainability of other options should be contemplated before the adaptive release approach is taken (Cassar and Pender 2005: 615).

Site characteristics to be considered

Site characteristics provide a telling narration of both historic use and maintenance, in addition to the likely success of any future interventions. Site location impacts material and plant availability, working hours (daylight and weather), tidal factors, as well as licence and permission requirements. Historically, a site's lack of accessibility may have been a function of its use, and public access was never a consideration. Today, remote and inaccessible assets are less likely to meet funding eligibility criteria, making private investment a necessity. The historic use of the site and its surrounding environment is also a factor, as long-term pollutant exposure may render sites unusable without decontamination. Modern environmental protections such as SSSIs (Sites of Special Scientific Interest) or SAC (Special Areas of Conservation) demand submission of Environmental Impact Assessments before works can be scheduled, contributing further administrative and financial costs to a project (Edwards 2023). The management of general environmental impacts and pollution, such as run off, erosion, and contaminated discharge, have to be considered as part of any project. The more challenging each of the above factors is to manage, the greater costs become, and the feasibility of conservation projects decreases. If these factors combine with a lack of maintenance, general disuse or public disinterest, and a lack of re-use potential, the less viable any higher-impact conservation measures are.

Using Multi-Attribute Value Theory in heritage decision making

Ferretti *et al.* (2014) outline examples where MCDA tools have been used to make decisions on cultural heritage management, principally in planning for the reuse of historic buildings. However, during the present study, the author found that in the last ten years, there has not been any research conducted into the use of MAVT (Fishburn 1967; Keeney and Raiffa 1976) specifically as a decision-making support tool in the context of conservation interventions for heritage at risk. Therefore, MAVT was selected as the main methodological approach to be trialled in this project.

Whilst the framework designed for this project could be used, with some adjustment, for other heritage types, it is significant to note that the presented MAVT methodology has been designed with listed buildings and intact scheduled monuments in mind. As approaches for archaeology, scheduled monuments and listed buildings could be different (Loader 2011: 106), the particularities of each site of interest should be taken into account prior to any application of the methodology.

MAVT analysis depends on value inputs from a range of stakeholders. Part of its merit is the capacity to use qualitative values, converted into quantitative values for scoring. In early stages of the project, it was determined that for the creation of a heritage at risk decision-making support tool that would be employable by heritage managers, it would need to be straightforward to use, with no baseline maths or statistical knowledge required, using largely pre-existing information and material. Moreover, the MAVT pathway to determining what can and cannot be lost, is too resource intensive to apply to all heritage at risk. For example, in Wales, 21% of listed buildings and 14% of scheduled monuments are currently at risk (Cadw 2023). Table 4.2 represents variations between maritime and industrial heritage at risk in Swansea, Carmarthenshire, and Wales as a whole. It is therefore necessary to screen heritage at risk, prior to engaging in the MAVT process, to determine which heritage assets would be good candidates.

Table 4.2: Heritage at Risk statistics, correlated from raw data provided by Cadw (2023). Copyright: Bella Romain.

	% of Maritime SMs	% of Maritime SMs at risk	% of Industrial SMs	% of Industrial SMs at risk	% of Welsh SMs	% of SMs at risk
Swansea	2.4%	67.0%	12.0%	67.0%	2.9%	26.0%
Carmarthenshire	0.5%	50.0%	60.0%	70.0%	8.7%	13.0%
Wales	0.4%	32.0%	60.0%	42.0%	100.0%	14.0%
	% of Maritime LBs	% of Maritime LBs at risk	% of Industrial LBs	% of Industrial LBs at risk	% of Welsh LBs	% of LBs at risk
Swansea	0.6%	67.0%	60.0%	39.0%	1.7%	22.0%
Carmarthenshire	0.3%	50.0%	42.0%	30.0%	6.1%	23.0%
Wales	0.6%	20.0%	17.0%	66.0%	100.0%	21.0%

For the presented project, the initial screening process, prior to the application of MAVT, was influenced by Swansea Council's Heritage at Risk Strategy (2013: 6), which prioritises assets by assigning unweighted values to attributes. The lower the total score, the higher the priority for intervention (Figure 4.1). Updated data from the risk register was used to score assets (Swansea Council 2016). Three buildings within Swansea, The Palace Theatre, Loughor Town Hall and Whitford Point Lighthouse, were chosen to evaluate an adapted screening tool which introduced weighting to various attributes. Table 4.3 explores these buildings and the conservation actions taken in recent history. Swansea Council's scores have been modified to correct errors or to reflect accurate condition. Subsequently, comparisons of the condition of the buildings between 2013 and 2019 were made to provide context for rates of deterioration in this period and how this was projected to continue. Table 4.4 demonstrates the screening exercise, with the introduced weightings for categories, applied to the three buildings. Data used in this type of exercise is readily available to heritage managers and requires no interpretation.

Heritage Status	Risk Category	Regeneration potential	Prominence
1 = Grade I	1 = Extreme risk	1 = High	1 = High
2 = Grade II*	2 = Grave risk	2 = Medium	2 = Medium
3 = Grade II	3 = At risk	3 = Low	3 = Low
	4 = To be watched		

Figure 4.1: Scoring matrix for heritage assets, adapted from Swansea Council 2013. Copyright: Bella Romain.

The most critical score the assessment can produce is 1.0, suggesting an urgent need for a decision on the heritage asset's future. When an asset is 'severely threatened and significant', MAVT assessment should be recommended (Dawson *et al.* 2021: 384). Without both, the asset is either not significant enough to warrant the investment for MAVT, nor under enough threat to prompt urgent action. This is because the purpose of the MAVT tool is to support decision making in circumstances where conservation is complex or challenging. The initial screening of the project identified both The Palace Theatre and Whitford Point Lighthouse as high priority for MAVT assessment because they are at extreme risk and restoration would be demanding. Whitford Point Lighthouse has the additional challenge of its intertidal location. Loughor Town Hall on the other had a lower priority because regeneration would be uncomplicated, therefore MAVT is not required to determine the feasibility of conservation choices and management plans.

Table 4.3: Heritage at Risk prioritisation exercise, adapted from Swansea Council, 2013. Copyright: Bella Romain.

Heritage Status	Risk Category	Regeneration Potential	Prominence
1 = Grade I	1 = Extreme Risk	1 = High	1 = High
2 = Grade II*	2 = Grave risk	2 = Medium	2 = Medium
3 = Grade II	3 = At risk	3 = Low	3 = Low
	4 = To be watched		

Year	Listed Building	Heritage Status	Risk Category	Regeneration Potential	Prominence	Total Score	Notes	Author's Notes
2013	Palace Theatre	3	1	1	1	**6**	Condition and feasibility study undertaken by the Design and Conservation Team. Currently in confidential discussions with the owner about the immediate condition issues and the longer term sustainable reuse potential. This building will require grant funding support to bring it back into use.	The Palace Theatre continued to decline with no apparent maintenance between 2015 and 2021. A SC led multi-million pound restoration began in 2021 to convert the building into offices. High status and regeneration potential resulted in restoration.
2016	Palace Theatre	3	3	1	1	**8**	Urgent works to remove vegetation and repair masonry and roof completed in 2015. Interior is badly affected by dry rot and water ingress. Ongoing discussions with owner to ensure sustainable reuse of the building. Level of risk reduced following completion of urgent works.	
2013	Loughor Town Hall	3	4	2	2	**11**	This building has recently been added to the Building at Risk Register. There is a need to undertake a condition and feasibility study and this could form the basis of funding bids working in partnership with the local community.	Regeneration potential should be at 1; there is nothing to prevent re-use of this building. No update to status since 2016. Community group proposal was still active in 2019, with fundraising efforts impacted by the pandemic. No further focused activity seems to have taken place. Low profile and medium regeneration potential has not resulted in any SC led efforts to re-use this building.
2016	Loughor Town Hall	3	4	2	2	**11**	Community group has proposals for sustainable reuse with community space at first floor brought back into use and flat on ground floor refurbished.	
2013	Whiteford Lighthouse	3	4	3	1	**11**	This is one of the iconic structures on Gower and discussions are needed with the owner (Carmarthenshire Council) about protection and consolidation.	Original strategy listed WPL as GII not GII*. According to SC's risk grading system, WPL should be at 1; the close scoring of The Palace and WPL with Loughor Town Hall masks the true risk to these buildings.
2016	Whiteford Lighthouse	2	3	4	1	**10**	Derelict. Also a Scheduled Ancient Monument. In exposed maritime location. No work since 2013.	
2019	Palace Theatre	3	2	1	1	**7**	Although urgent works have been carried out, the building as a whole is in a bad state of repair. Talks have been underway for a long time, but nothing has happened to date. Risk reduced to 2, as the building is still vacant and aspects of internal failure cannot be ruled out.	These scores have been assigned by the author to represent the true risk to the HAs. All assets are considered as they were in 2019, prior to any works being carried out. This scoring forms the basis of the prioritisation assessment.
	Loughor Town Hall	3	4	1	3	**11**	A CMP was drawn up for the building in the last two years, with a view to reuse as a community building. Currently there are no further plans, but the building is stable, although vacant. It is important to the local community but its prominence does not spread far beyond that.	
	Whiteford Lighthouse	2	1	4	1	**8**	The risk of structural failure to WPL cannot be ignored as the skirt continues to be eroded and undermined. The complete loss of the roof covering has opened the lantern and store-rooms to the elements, with a high likelihood of water ingress into the main structure. This is highly likely to cause corrosion internally, and saturation of the mortar could lead to instability.	

Subsequently, after the identification of the relevant heritage assets that would benefit from further assistance in the conservation decision making, the MAVT process can be conducted in the following steps:

1. The objective, or question, of the MAVT is decided upon.
2. Attributes are considered. These are factors which affect the desirability or success of each alternative.
3. Define alternatives.
4. All experts (from the various stakeholders participating in the MAVT survey) weight each attribute for how important it is in the consideration of the alternatives. Some weights will reflect on the importance of the attribute (for example, how much significance does the environmental protection have?), others will consider the impact of the attribute on the viability of the alternative (for example, to what extent is long-term cost of maintenance a factor benefiting or disadvantaging conservation?).
5. Calculate a mean average for the weight of each attribute.
6. Determine the qualitative values for each attribute.
7. Assign quantitative value functions to qualitative values.
8. Score each alternative against the attributes, first with a qualitative value, and then translated to a quantitative.
9. Calculate the weighted score for each alternative.
10. Aggregate the total weighted score of each alternative.
11. Rank the alternatives.
12. Sensitivity test for the alternatives.

Each alternative is scored according to the following equation:

$$\mathbf{V(a) = \Sigma w_i \cdot v_i(a_i)}$$

Where $\mathbf{V(a)}$ is the total score of 'alternative' **a**, $\mathbf{w_i}$ represents the weight of attribute **i** and $\mathbf{v_i(a_i)}$ represents the value function of 'alternative' **a** in consideration of attribute **i**, as also described by Ferretti *et al.* (2014).

Table 4.4: Heritage at Risk prioritisation assessment. Copyright: Bella Romain.

Weighting & Scoring																		
Weights →		Status	0.15	Weighted Score	Risk	0.25	Weighted Score	Rarity	0.2	Weighted Score	Prominence	0.2	Weighted Score	Development potential	0.2	Weighted Score	Totals	
Scores →		GI/ SM	1		Extreme	1		International	1		International	1		High	1			
		GII*	0.75		Grave	0.75		National	0.75		National	0.75		Medium	0.67			
		GII	0.5		At risk	0.5		Regional	0.5		Regional	0.5		Low	0.33			
		Local List/ CA	0.25		To watch	0.25		Local	0.25		Local	0.25		None	0			
↓ Assets								None	0		None	0						Notes
Palace Theatre	At extreme risk, prominent within Wales and nationally rare. There is a high re-use prospect due to its location and the organisation of the building. 	GII	0.50	0.075	Grave	0.750	0.188	Regional	0.500	0.100	National	0.750	0.150	High	1.000	0.200	**0.713**	In this example, five pieces of readily available information have been used to prioritise assets for MAVT assessment. The higher the score, the greater the priority; if either significance or threat is high, the MAVT assessment process should be triggered, with a view to assessing the options for its future. This process has identified both the Palace Theatre and Whitford Point as high priority for assessment because they are at extreme risk and restoration would be intensive. Whitford Point has the additional challenge of its location. Loughor Town Hall is a lower priority because there is less work involved in restoration and redevelopment, and therefore MAVT is not required to determine the viability.
Loughor Town Hall	Significance is mainly local historical value. SC describe it as having medium regeneration potential but it has been used for residential and commercial so redevelopment should be straightforward. 	GII	0.50	0.075	To watch	0.250	0.063	None	0.000	0.000	Local	0.250	0.050	High	1.000	0.200	**0.388**	
Whitford Point Lighthouse	Cast-iron lighthouse in the ITZ of the Loughor Estuary. Highly significant and of international importance; one of only two in the world. Due to its location and condition there is no reuse potential. 	GII*	0.750	0.113	Extreme	1.000	0.250	International	1.000	0.200	International	1.000	0.200	None	0.000	0.000	**0.763**	

Sensitivity analysis is then performed to determine which attributes have the greatest impact on the score. This involves adjusting the weight of the attributes so that one is set to high while the others are low, repeating the process for each attribute in turn. In the current project, the first sensitivity test (S_1) was based on that used by Ferretti *et al.* (2014), but these weights were too narrow for the large number of attributes in this study. Therefore, a second sensitivity test (S_2) was conducted, which increased the range between the higher weight and the control weights.

Application of Multi-Attribute Value Theory to Whitford Point Lighthouse

This case study considers the significance of Whitford Point Lighthouse, the risks and challenges it faces, and the conservation interventions available to it, before discussing these factors with a range of experts and submitting Whitford Point Lighthouse to the MAVT exercise to assess its usefulness as a decision-making support tool for complex heritage at risk cases, where loss is expected.

Significance

In the 19th century, cast-iron was increasingly used as a construction material due to its compressive strength and ability to survive harsh environments (Van Dyke 2004: 27). Its prolific use in lighthouses is attributed to the lighter weight, lower cost, and quicker fabrication and assembly speeds when compared to masonry (Lewis 2012). Alexander Gordon, a British engineer, was responsible for the design and distribution of many cast-iron tower lighthouses around the world.

The first known cast-iron lighthouse was built in 1803 on Swansea's West Pier (Coflein 2009) and formed one of a cluster of cast-iron lighthouses built in South Wales. Most wave-washed lighthouses were masonry built, but estuaries were unable to support the weight, leading to the use of piled lighthouses (Eshelman 1997: 6). A lighthouse was necessary in the Loughor Estuary due to the shifting sandbanks which could easily ground a ship. In 1865, after a series of storms and collisions destroyed the original timber lighthouse (Hartley 2003: 31), the Llanelly Harbour Commissioners decided to erect an iron lighthouse in the estuary. The lighthouse was designed by John Bowen, Llanelly Copperworks' chief engineer, and built by Hennet, Spink and Else of Bridgwater. It was completed, more than a year overdue, in October 1866 (Hartley 2005: 20). The Whitford Point Lighthouse (Figure 4.2) required extensive interventions in its early years, but by 1885 it was stable, and no further deterioration occurred (Figure 4.3). However, on December 16th, 1920, Whitford Point Lighthouse was decommissioned, and all its fixtures removed (Carmarthenshire Archive Service n.d.). The lighthouse has received no maintenance since, barring some repairs to the concrete skirt around the base in the late 20th century. The structure is currently undermined by at least one foot around more than 30% of the base (Figure 4.4).

The only other cast-iron wave-washed tower lighthouse is the Roter Sand Lighthouse (Lewis 2023), built in 1885 *c.*11 miles from Bremerhaven, in Germany, on a caisson sunk onto the seabed. In 1987 it became a national monument and was restored (Deutsche Stiftung Denkmalschutz 2023).

Figure 4.2: Whitford Point Lighthouse. Copyright: Bella Romain.

Figure 4.3: Wrought-iron straps on Whitford Point Lighthouse. Copyright: Bella Romain.

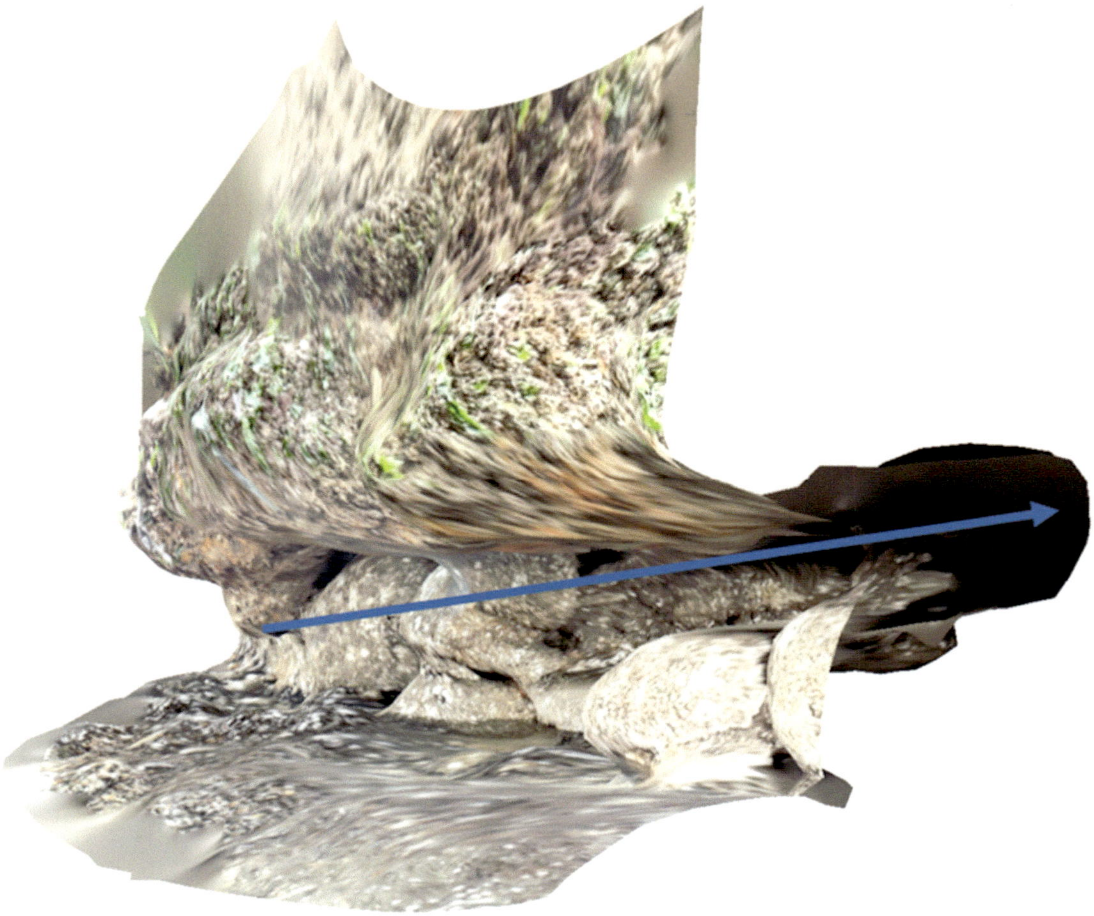

Figure 4.4: Undermining of the base, 3D capture. Copyright: Bella Romain.

The significance of the Whitford Point Lighthouse has the following values:

Historic

The Whitford Point Lighthouse is associated with many people of note through its construction and design. It is of exceptional importance and incredibly rare. Its significance is somewhat reduced by the removal of its components upon decommissioning (ICOMOS and TICCIH 2011: A9).

Evidential

The Whitford Point Lighthouse provides a unique insight into the materials and techniques of its build. It also provides information on the behaviour of historic cast-iron in seawater. Documentary evidence of its early maintenance provides insight into the efficacy of these methods. It is the only plate lighthouse known to be bolted externally, believed to be for ease of swift erection and plate removal due to its tidal location.

Aesthetic

A structural survey (Nicholas 1997: 5) suggested that the Whitford Point Lighthouse was 'at risk of losing its design legibility' due to the ongoing erosion of its features. Its iconic shape and silhouette are the subject of art, photography, and branding (see Romain 2023: 38, figure 16-18).

Communal

The Whitford Point Lighthouse is a beloved seamark in the area and is visible from the shore all around the estuary (Figure 4.5). It received renewed and widespread attention in 2022 when it was lit for the Queen's Jubilee.

Figure 4.5: Zone of Theoretical Visibility map of Whitford Point Lighthouse. Copyright: Bella Romain.

Risks, Challenges, and Limitations

Many of the challenges in conserving Whitford Point Lighthouse are cumulative of its situation. Over a century in the intertidal zone has left the iron saturated in salt water, corroding, and covered in biofouling. The bricks of the living quarters are similarly showered in salt and guano. Any attempts to protect the metal work with a fresh coat of paint would be futile, as the trapped salts would promote corrosion (Watkinson 2022: 14). The structure would need desalinating prior to any works, which would require disassembly and transportation of the lighthouse. Its location in the intertidal zone restricts the *in situ* working period to a few hours a day and complicates the prospects of transportation via a barge. Additionally, the lighthouse is only accessible by an off-road vehicle with permission from neighbouring landowner, the National Trust. Environmental protection of all the land, marsh and sea around Whitford Point elicit the need for licences and permissions from Natural Resources Wales. Decommissioning and the subsequent lack of maintenance over the succeeding century has contributed to a substantial loss of original features. Its future is at risk and although its significance and rarity make it worthy of saving, the measures needed to secure its future are extreme and costly. Here are all the risk, challenges and limitations in detail:

Location

The Whitford Point Lighthouse is situated in the Loughor Estuary, lined with treacherous sandbanks. The Bristol Channel has the second highest tidal range in the world, which affects current speed and safety at the site. The surrounding area has several protective ecological designations and is difficult to access. The heavy industrial past of the estuary and the lighthouse's location in the intertidal zone has resulted in considerable localised areas of decay and corrosion, particularly to wrought-iron elements, such as the flange bolts (Figure 4.6).

Figure 4.6: Corrosion of wrought-iron bolts. Copyright: Bella Romain.

Environmental

The numerous conservation designations (Table 4.5) have an impact on activities and licencing requirements, managed by Natural Resources Wales. Any work carried out on the Whitford Point Lighthouse would be subject to approval following submission of an Environmental Impact Assessment, contributing a significant financial burden to any proposed strategy (Edwards 2023). Working in an enclosure on the Whitford Point Lighthouse is not possible due to its location in the intertidal zone. Even if desalination *in situ* was possible, containment could not be guaranteed and the designations of the site and potential for environmental harm caused by cleaning would prohibit this (Cadw 2020: 28).

Table 4.5: Legislation and Designation. This table outlines acts, policies, and guidance that apply to the site. Copyright: Bella Romain.

The table below outlines the acts, policies and guidance documents that apply to the site				
	Statutory	Policy	Guidance	Nature Designations
EU*	EU Marine Area			RAMSAR (Wetlands of International Importance) SAC (Special Areas of Conservation) SPA (Special Protection Area)
UK	Marine and Coastal Access Act 2009 The Planning (Listed Buildings and Conservation Areas) Act 1990 The Ancient Monuments and Archaeological Areas Act 1979	Marine Policy Statement (UK) 2021		SSSI (Site of Special Scientific Interest) AONB (Area of Outstanding Natural Beauty) Heritage Coast National Nature Reserve Local Nature Reserve
Wales	Historic Environment (Wales) Act 2023 The Historic Environment (Wales) Act 2016 The Planning (Listed Buildings and Conservation Areas) (Wales) Regulations 2012	Planning Policy Wales Technical Advice Notes Supplementary Planning Guidance Wales Marine Plan	Lavernock Point to St Ann's Head SMP2 (Shoreline Management Plan) Seascape Assessment Caring for Coastal Heritage Managing the Marine Historic Environment of Wales Historic Landscape Character Assessment Marine Character Assessment	

* Although the UK is no longer part of the EU, these protections were formally adopted by the UK and have been maintained following the UK's exit from the EU.

Administrative

The Whitford Point Lighthouse was built by the Llanelly Harbour Commissioners. Responsibility later fell to the Llanelly Harbour Trust (Llanelly Harbour Act 1896) which dissolved in 1997, and all assets transferred to Carmarthenshire County Council (The Llanelli Harbour Revision Order 1997). Responsibility for the oversight of listed buildings and scheduled monuments falls to the local authority, as defined by geographical boundaries. Whitford Point Lighthouse, despite being owned by Carmarthenshire County Council, sits within the borders of Swansea Council. Several factors inhibit clear recording of heritage assets, including the duplicated submission of individual sites onto the historic environment record (Whitewright 2023) and a lack of joined-up thinking between council departments for management of out-of-county property (Hellier 2023). The dual designation of Whitford Point, as both a listed building and a scheduled monument, creates a contradiction of conservation objectives (Smith 2023). Whilst it enjoys greater theoretical protection, its remote location ensures that surveillance is minimal, a fact confirmed by the unauthorised installation of lights on the lantern balcony for the Jubilee.

Logistics

Considerations such as site access, groundworks, and transportation are liable to increase the costs of any construction project significantly (Cunningham 2013: 10). Oil and Gas UK (2013: 118) found that inshore and offshore works cost approximately 30 and 150 times more, respectively, than the same work carried out on shore. In 2008, Acanthus Holden (2008: 18) estimated the costs to restore Whitford Point Lighthouse at £875,543 (adjusted for inflation according to the Bank of England; calculated in 2022). This did not account for total disassembly of the structure to allow for cleaning, nor for the environmental implications of otherwise doing the works *in situ*. The Whitford Point Lighthouse's location exacerbates logistical difficulties, as access, water, power, and plant delivery are not straightforward (Kaltenbacher 1997: 12). Access to the Whitford Point Lighthouse is limited as there is no suitable vehicular access, and the intertidal zone restricts access according to tides, weather, and season. Furthermore, the sandbanks complicate access by sea. Hence, the Whitford Point Lighthouse cannot offset its maintenance costs because it has no potential commercial use. Douglas Hague (1972: 46) lamented that 'such a building cannot be maintained unless it is in use'.

Anthropogenic Climate Change

In Wales over 50% of heritage assets (including archaeological sites) are at immediate or short-term risk from erosion (Davidson 2002: 117; Wienberg 2014: 71) and approximately 16,000 face loss due to sea-level rise (Historic Environment Group 2020: 28), while maritime industrial heritage is at highest risk (Cook 2019: 169). The pace of climate change means loss of coastal heritage is inevitable, no matter what measures are taken (DeSilvey and Harrison 2019: 2; Historic Environment Group 2020: 18; Venture *et al.* 2021: 423). Associated risks include more frequent storms, sea-level rise, erosion, and elevated acidity of seawater (Casey and Becker 2019: 170; Hassan and Xie 2020: 18; Murphy *et al.* 2009: 11).

The greatest impact on the lifespan of the Whitford Point Lighthouse is likely to be the effects of sea-level rise and the combined action of increased storm surges and erosion which will leave Whitford Spit and Whitford Point vulnerable. Increased corrosion caused by acidic seawater is unlikely to have a pronounced effect on its own, but could weaken materials, particularly bolts, resulting in structural failure from other factors. Predictions for sea-level rise in Wales vary between 0.27m and 1.13m by 2100 (Hayles 2022: 18). Catastrophic storms in 2014 saw adjacent Cwm Ivy's medieval seawall, which protected acres of freshwater marsh, breached. It is anticipated that the Whitford

Point Lighthouse will see an increase in storm surge heights of approximate 1 meter (Lowe and Gregory 2005: 1317) and that these events will be of greater consequence than sea level rise alone (Lowe *et al.* 2001: 179).

Risk Assessment

A detailed assessment of all the relevant risks was conducted and is presented in Table 4.6.

Table 4.6: Risk assessment of Whitford Point Lighthouse. Copyright: Bella Romain.

Risk	What	Chance	Harm	Score	Proposal
Environmental factors	Historic pollution accelerating corrosion processes.	LOW	LOW	1	Majority of factors no longer contribute to the issue; no action required.
Logistics	Inaccessible for works to be carried out- high costs involved, routine work less likely to be done.	HIGH	HIGH	9	Access plan needs to be developed if WPL is not moved from setting. Regular checks of structure. Translocation of WPL.
Designations	Environmental designations create administrative burden and delays to works.	HIGH	LOW	3	Environmental Impact Assessments could be drawn up in advance to ensure minimal delays to urgent works.
Maintenance	Lack of maintenance means WPL would need dismantling before any works or corrosion would accelerate	HIGH	HIGH	9	Historic lack of maintenance is responsible for current state. Cost of reversal is high. Restoration is the only option to mitigate.
Erosion	Erosion of the plinth may undermine WPL's stability; erosion of Whitford Spit could lead to increased storm surges.	HIGH	MED	6	Erosion of plinth is ongoing. Regular maintenance and renewal would reinforce stability. Whitford Spit could be protected through dredging of Ebb Delta sandbank. Translocation of WPL, or demolition as last resort.
Sea Level Rise	Sea level rise with storm surges could result in more mechanical damage to WPL. Sea level rise likely to lead to greater erosion of Whitford Spit.	MED	MED	4	No way to mitigate against sea level rise; ensuring structure is sound is the best way to minimise damage. Translocation of WPL or demolition if harm to public likely.
Storm surges	Variance in bathymetry could result in stronger waves striking WPL.	MED	MED	4	As above.
Increased acidity	Accelerated corrosion of metal, particularly wrought iron bolts holding structure together.	MED	LOW	2	Ensuring dissimilar metals are insulated from each other, and replacement of selected bolts in structure would ensure WPL was sound.

Interventions

Lighthouse specific conservation literature is limited to mainly USA-based case studies and maintenance guides. Although governance varies, these documents provide a comprehensive review of interventions to lighthouses, including those constructed with iron (e.g., Aument and Wathne 2012; MacDonald 2018; Wallace 2009). The options for conservation of the Whitford Point Lighthouse considered in this paper have previously been suggested by both Faulkner (2022: 19) and Acanthus Holden (2008). Several investigations were conducted between 1997 and 2008 (Acanthus Holden 2008; Howe 1997; Nicholas 1997) which provide details regarding costings and feasibility, and these have informed the MAVT process. The Deutsche Stiftung Denkmalschutz (2023: 7) also proposed these interventions for the Roter Sand Lighthouse in Germany (built in 1885 and in use until 1964), which was restored and used for tourist accommodation, generating revenue for maintenance, but has since been closed due to worsening condition.

Restoration

No plans or drawings of the Whitford Point Lighthouse survive to inform future restoration works.

Translocation

The first known translocation as a means of conservation was carried out in the UK in 1598 (Drozd 2019: 177). Erosion is the motivation behind most lighthouse relocations (DeBlieu and Halminski 2001; Wienberg 2014) and the first UK lighthouse to move was Smeaton's Tower (Figure 4.7) on Plymouth Hoe which was disassembled before transportation in 1882 (Bayley 2017). Translocation of the Whitford Point Lighthouse is impossible by land, and complex and costly by sea. The difficulties of finding a new site and the logistics of working in the intertidal zone make translocation an unlikely option.

Figure 4.7: Smeaton's Tower, Plymouth Hoe. Copyright: Bella Romain.

Adaptive Release[1]

Adaptive release was an option considered for the Roter Sand Lighthouse. Although experts recognised the benefits of managed decline, the lack of public access, environmental impact, and the inevitable total loss were considered too great a price (Deutsche Stiftung Denkmalschutz 2023: 11). The Wyre Light in Lancashire is currently in a perilous state and little of the original structure remains (Hunt 2023). Despite near collapse, heritage values are preserved through community engagement activities, following the principles of adaptive release. Improvements in technologies used for digital humanities have resulted in techniques which offer immersive virtual experiences (for example, 3D walkthroughs, reconstructions, 360° views) and opportunities to preserve buildings and artefacts digitally for posterity, for example through 3D recording and laser scanning (Figure 4.8). These methods not only create a permanent record, but enhance access for those with mobility issues, providing learning and research possibilities irrespective of where someone is in the world, and developing a resource that can be used for a multitude of purposes.

[1] Also described as 'planned release' in various parts, graphs and figures of the research project.

Figure 4.8: 3D capture of Whitford Point Lighthouse in 2023. Copyright: RCAHMW, 2023 (open access source available at Sketchfab).

Demolition

Demolition is usually a last resort, proactively taken in the name of safety or salvage. An example of such a conservation choice is the Orfordness Lighthouse, which was demolished to prevent uncontrolled loss and to salvage historically significant elements (Barber and Cawley 2020).

Attributes

The range of site-specific intervention attributes has to be adapted to each asset and its landscape. The variables that exist for any given project are unique to each situation and should be addressed as such. Attributes which should be considered for any use of the MAVT tool include the impact to its significance, accessibility, environmental factors, budget constraints, the possibility of further funding, and re-use potential. The degree to which each of these is important will vary and is to be weighted by relevant experts. Other attributes might include availability of materials or skilled craftspeople, longevity of the intervention, and the impact on other assets.

The attributes that were considered for the MAVT tool of the Whitford Point Lighthouse are presented in Table 4.7. 'X' refers to any one of the alternatives.

Table 4.7: Attributes considered for the MAVT tool of the Whitford Point Lighthouse. Copyright: Bella Romain.

Attribute		Justification
Values	Historic	How will the values and significance of the lighthouse be affected if WPL is 'X'?
	Evidential	
	Aesthetic	
	Communal	
Opportunity for Engagement		What are the opportunities for communities and members of the public to engage with the process of 'X', or afterwards?
Accessibility		How does 'X' improve accessibility for people with mobility needs or other disabilities?
Short Term Costs		What are the initial costs of 'X'?
Long Term Costs		What are the ongoing maintenance costs of 'X'?
Funding Opportunities		Is funding available for 'X'? Will it be easy to access?
Risks to HE / Archaeology		What risks are there to the historic environment or other archaeology in the landscape (excluding WPL) with 'X'?
Risks to Environment		What risks are there to the natural environment and wider landscape with 'X'?
Impact on Authenticity		How does 'X' impact the authenticity of WPL? Can it be mitigated?
Logistical Possibility		How viable is 'X'? Do the logistical difficulties make it prohibitively expensive?

Multi-Attribute Value Theory for Whitford Point Lighthouse

As part of this project and case study investigation, experts were identified as people involved in managing heritage assets or the coastline at Whitford Spit (Davidson 2002: 119), and specialists who contributed specific knowledge regarding the logistics and cost of the conservation alternatives (Table 4.8). The experts were given a survey questionnaire (processed anonymously) to weigh each identified attribute according to how much they felt it should influence the scoring of conservation alternatives (Table 4.9). Weightings are continuous and attributes can share the same weight, but the total weight of all attributes must add up to 1.0 (100%). A mean average was calculated from the accumulated weighting. Weighting in this way reflects the perceived overall importance of the attribute.

Table 4.8: Roles of experts approached for the project and their familiarity with Whitford Point Lighthouse. Copyright: Bella Romain.

Expert	Role	Familiarity with Whitford Point Lighthouse
i	Conservation Officer	Familiar
ii	Conservation Structural Engineer	Aware
iii	Marine Civil Engineer	Aware
iv	Maritime Archaeologist	Familiar
v	Heritage Consultant	Very
vi	Environmental Consultant	Very

Table 4.9: Attribute weighting by experts. Copyright: Bella Romain.

Weighting						
Expert → / Alternative ↓	i	ii	iii	iv	v	Average weighting
Historic	0.1	0.1	0.07	0.1	0.08	0.090
Evidential	0.05	0.04	0.02	0.075	0.07	0.051
Aesthetic	0.1	0.09	0.07	0.075	0.08	0.083
Communal	0.1	0.04	0.07	0.075	0.07	0.071
Opportunity for Engagement	0.1	0.04	0.05	0.075	0.07	0.067
Accessibility	0.05	0.05	0.05	0.1	0.06	0.062
Short Term Costs	0.05	0.08	0.1	0.1	0.12	0.090
Long Term Costs	0.05	0.08	0.12	0.05	0.07	0.074
Funding Opportunities	0.1	0.04	0.13	0.05	0.1	0.084
Risks to HE/ Archaeology	0.05	0.1	0.06	0.05	0.05	0.062
Risks to Environment	0.1	0.12	0.1	0.05	0.08	0.090
Impact on Authenticity	0.05	0.08	0.03	0.1	0.06	0.064
Logistical Possibility	0.1	0.14	0.13	0.1	0.09	0.112
Total	1	1	1	1	1	1

Subsequently, experts were requested to score each alternative based on their own specialist area. For example, experts ii and iii scored the logistical viability of each option, using the qualitative values indicated in Table 4.10.

Table 4.10: Attribute / Alternative qualitative values. Copyright: Bella Romain.

Attribute	Qualitative Value				
Value Impacts	Bad	Discrete	Good	Very Good	
Opportunities for Engagement	None	Desk based	Recording	Interpretation	Heritage skills
Accessibility	No disabled access	In setting	In museum		
Short Term Costs	>£3 million	£1-3 million	<£1 million		
Long Term Costs (per 5 yr.)	£20,001<£30,000	£10,001<£20,000	<£10,000		
Funding Opportunities	Poor	Average	Good		
Risks to Archaeology	Severe	High	Some	Minimal	
Risks to Environment	Severe	High	Some	Minimal	
Impact on Authenticity	Strong negative	Negative	Discrete	Positive	Strong positive
Logistical Possibility	Extremely complex	Difficult	Simple		

The qualitative scores were assigned to the alternatives (Table 4.11) before being converted into quantitative value functions between 0 and 1.0 (Tables 4.12a, 4.12b and 4.13).

Table 4.11: Alternatives - qualitative values. Copyright: Bella Romain.

Attribute		Alternatives					
		Minimal Intervention	Restoration	Relocation within setting	Relocation to OAM	Planned Release	Demolition
Impact on Values	Historic	Discrete	Good	Discrete	Discrete	Discrete	Bad
	Evidential	Discrete	Good	Good	Bad	Discrete	Bad
	Aesthetic	Discrete	Very Good	Good	Good	Discrete	Bad
	Communal	Discrete	Very Good	Good	Very Good	Good	Bad
Opportunity for Engagement		Interpretation	Heritage Skills	Heritage Skills	Heritage Skills	Interpretation	Recording
Accessibility		No disabled access	No disabled access	In Setting	In Museum	No disabled access	No disabled access
Short Term Costs		<£1m	£1-3m	>£3m	>£3m	<£1m	£1-3m
Long Term Costs		<£10000	£20,001<£30,000	£10,001<£20,000	£10,001<£20,000	<£10000	<£10000
Funding Opportunities		Poor	Good	Good	Average	Average	Poor
Risks to HE/ Archaeology		Minimal	Minimal	Some	High	Some	Some
Risks to Environment		Minimal	Minimal	Severe	Some	Some	Severe
Impact on Authenticity		Discrete	Negative	Discrete	Negative	Discrete	Strong negative
Logistical Possibility		Simple	Extremely complex	Extremely complex	Extremely Complex	Simple	Difficult

Table 4.12a: Table demonstrating the quantitative value functions assigned to qualitative scores. 1.0 is the highest score and 0 is the lowest. Copyright: Bella Romain.

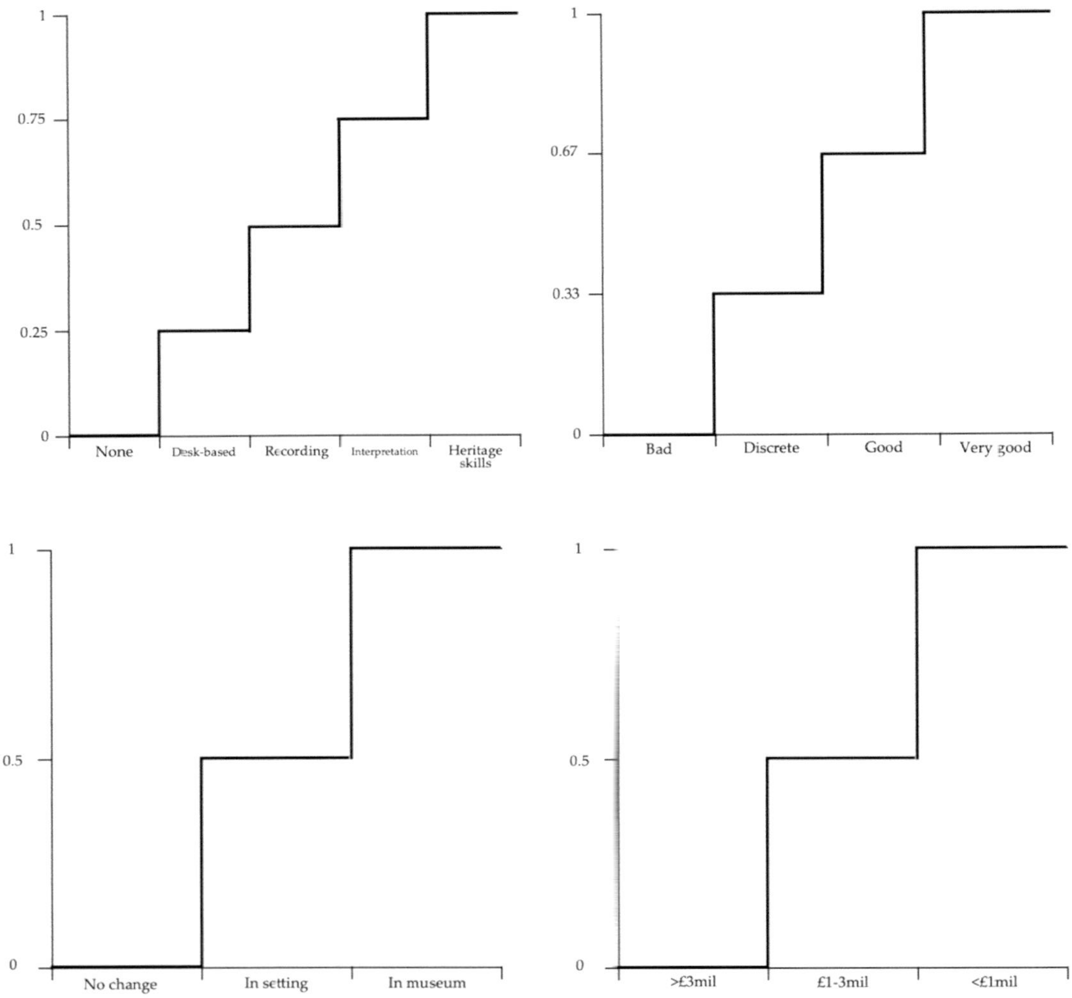

Table 4.12b: Table demonstrating the quantitative value functions assigned to qualitative scores. 1.0 is the highest score and 0 is the lowest. Copyright: Bella Romain.

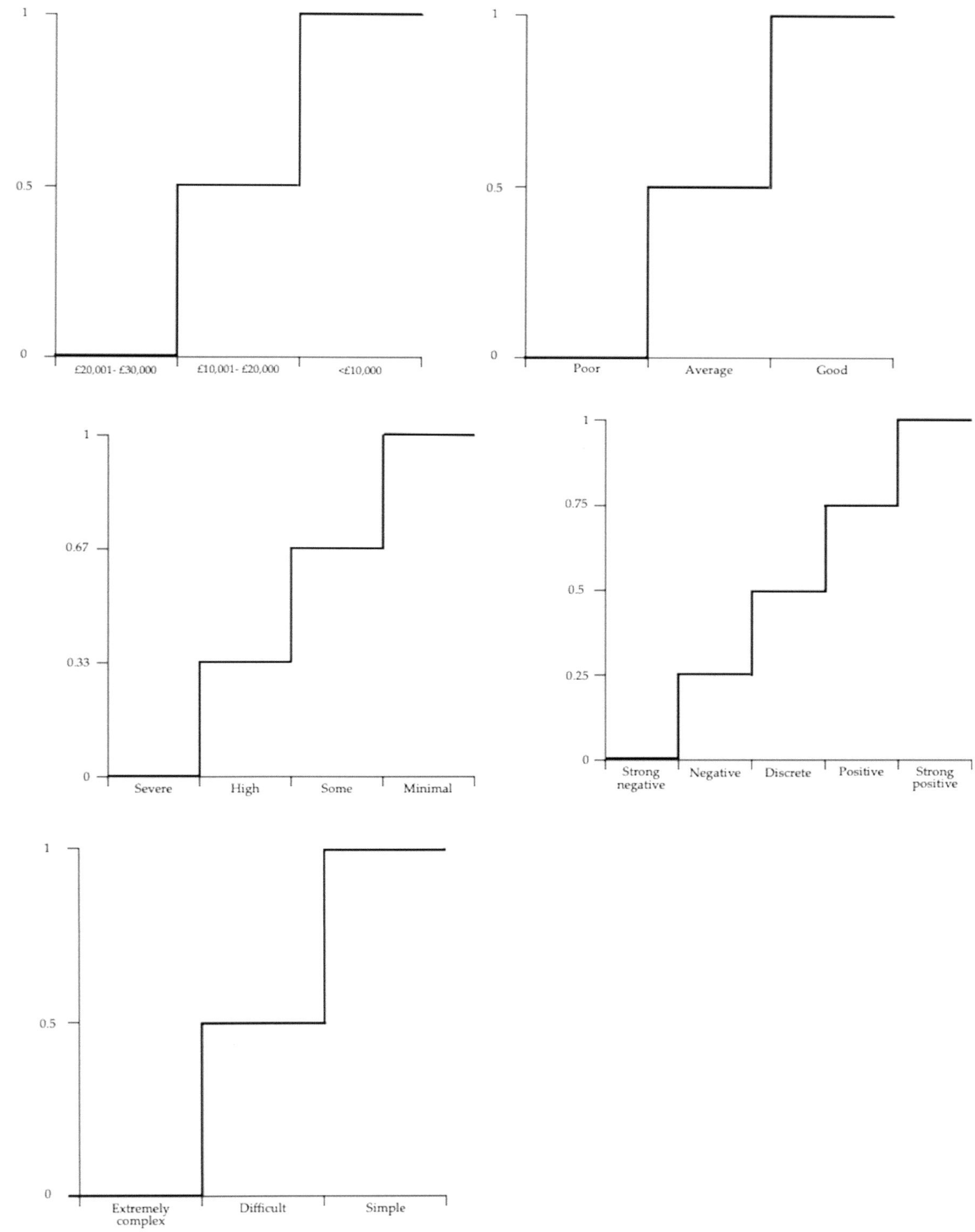

Table 4.13: Alternatives - quantitative values. Copyright: Bella Romain.

Attribute		Alternatives- Quantitative					
		Minimal Intervention	Restoration	Relocation within setting	Relocation to OAM	Planned Release	Demolition
Impact on Values	Historic	0.33	0.67	0.33	0.33	0.33	[illegible]
	Evidential	0.33	0.67	0.67	0	0.33	[illegible]
	Aesthetic	0.33	1	0.67	0.67	0.33	[illegible]
	Communal	0.33	1	0.67	1	0.67	[illegible]
Opportunity for Engagement		0.75	1	1	1	0.75	0.5
Accessibility		0	0	0.5	1	0	[illegible]
Short Term Costs		1	0.5	0	0	1	0.5
Long Term Costs		1	0	0.5	0.5	1	[illegible]
Funding Opportunities		0	1	1	0.5	0.5	[illegible]
Risks to HE/ Archaeology		1	1	0.67	0.33	0.67	0.67
Risks to Environment		1	1	0.25	0.67	0.67	[illegible]
Impact on Authenticity		0.5	0.25	0.5	0.25	0.5	[illegible]
Logistical Possibility		1	0	0	0	1	0.5

Quantitative values were multiplied by the weighted attribute values to calculate a score for each alternative (Table 4.14). The total score for each alternative determines its ranking, creating an expertly judged set of alternatives. The closer to 1.0 the score, the higher the rank.

Table 4.14: Scores of heritage management options – $w_i \cdot v_i(a_i)$. Copyright: Bella Romain.

Attribute		Weight	Minimal Intervention	Restoration	Relocation within setting	Relocation to OAM	Planned Release	Demolition
Impact on Values	Historic	0.0900	0.0297	0.0603	0.0297	0.0297	0.0297	0.0000
	Evidential	0.0510	0.0168	0.0342	0.0342	0.0000	0.0168	0.0000
	Aesthetic	0.0830	0.0274	0.0830	0.0556	0.0556	0.0274	0.0000
	Communal	0.0710	0.0234	0.0710	0.0476	0.0710	0.0476	0.0000
Opportunity for Engagement		0.0670	0.0503	0.0670	0.0670	0.0670	0.0503	0.0335
Accessibility		0.0620	0.0000	0.0000	0.0310	0.0620	0.0000	0.0000
Short Term Costs		0.0900	0.0900	0.0450	0.0000	0.0000	0.0900	0.0450
Funding		0.0740	0.0740	0.0000	0.0370	0.0370	0.0740	0.0740
Long Term Costs		0.0840	0.0000	0.0840	0.0840	0.0420	0.0420	0.0000
Risks to HE/ Archaeology		0.0620	0.0620	0.0620	0.0415	0.0205	0.0415	0.0415
Risks to Environment		0.0900	0.0900	0.0900	0.0000	0.0603	0.0603	0.0000
Impact on Authenticity		0.0640	0.0320	0.0000	0.0320	0.0000	0.0320	0.0000
Logistical Possibility		0.1120	0.1120	0.0000	0.0000	0.0000	0.1120	0.0560
		TOTALS	**0.6076**	**0.5965**	**0.4596**	**0.4451**	**0.6236**	**0.2500**

Following that, sensitivity testing was performed to determine whether the calculated outcome was true regardless of which attributes were afforded a higher weighting. The logic behind this is that small nuances in weighting could have a magnified effect on the outcome of the 'preferred' alternative. By running the analysis again, this time with each attribute taking its turn as the definitive weight, outliers are visualised and trends in data become apparent.

Discussion of Results

Weighting

The attribute weighting indicates how experts perceived the importance of each attribute (see Table 4.9). Figure 4.9 shows the mean weighting of the attributes alongside the maximum and minimum weights provided by experts. Closely weighted attributes are indicated by shorter high-low bars on the chart. The closer the mean line is to the centre of the high-low bar, the more evenly distributed the weightings. Where the mean line is near an extreme of a bar, it indicates that the opposite end of the bar is an outlier, and the data trends towards the mean weight. Agreement between experts was not expected due to their different professions and priorities. This was an advantage as it meant the MAVT was more authentic and had genuine conflicting opinions.

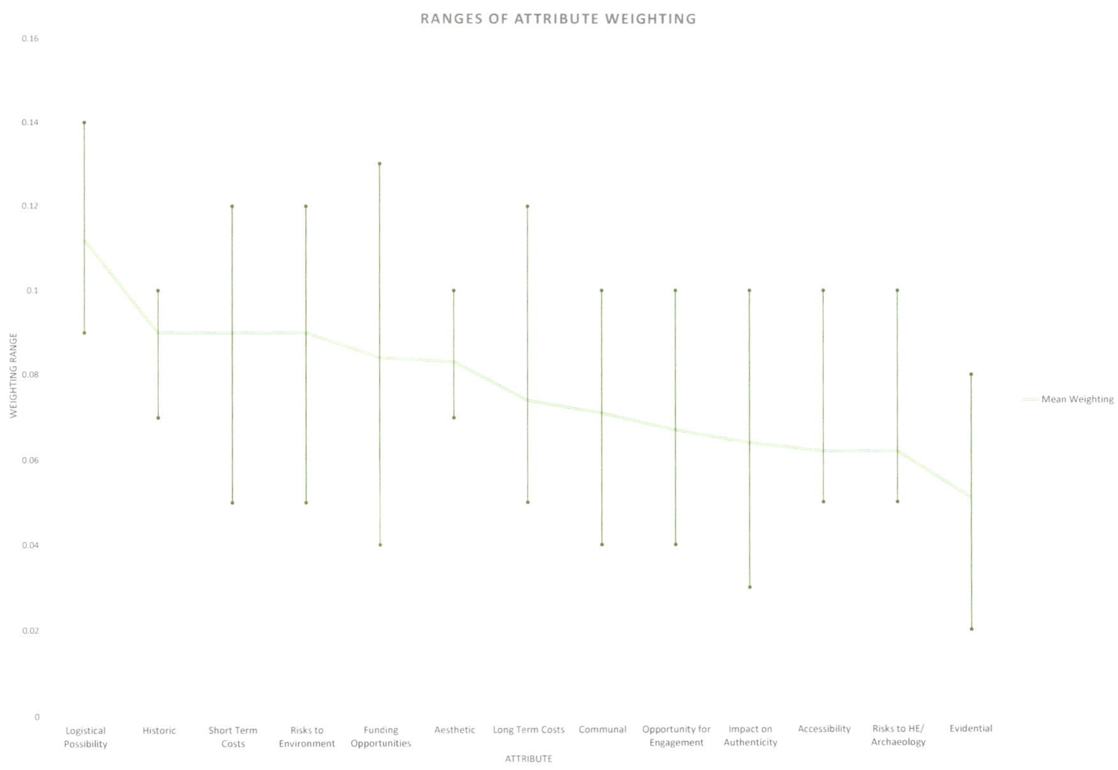

Figure 4.9: Graph showing attribute weighting ranges. Copyright: Bella Romain.

The range of weights (Figure 4.9 and Table 4.15) indicates how congruent experts' opinions were. The narrowest ranges were the historic and aesthetic values, indicating that all professionals felt similarly about the importance of these attributes. The most divergent attribute was funding opportunities. Given the nature of the Whitford Point Lighthouse, and the associated costs with any of the alternative conservation options, this was unexpected. Constraints proved to be the highest weighted attributes. Logistical possibility was deemed to be the most important, representing the fact that Whitford Point Lighthouse's location is prohibitive of most interventions (Smith 2023), and the practicalities of any alternative must be given thoughtful consideration. Initial costs and environmental risks also ranked high in the weighting. The median attribute was long-term cost, reflecting the need to assess the sustainability of any proposal, and equally, to avoid any future financial liability (Emerick 2014: 194).

Table 4.15: Weighting ranges showing how experts perceived the importance of each attribute. Copyright: Bella Romain.

	Logistical Possibility	Historic	Short Term Costs	Risks to Environment	Funding Opportunities	Aesthetic	Long Term Costs	Communal	Opportunity for Engagement	Impact on Authenticity	Accessibility	Risks to HE/ Archaeology	Evidential
Minimum	0.090	0.070	0.050	0.050	0.040	0.070	0.050	0.040	0.040	0.030	0.050	0.050	0.020
Mean	0.112	0.090	0.090	0.090	0.084	0.083	0.074	0.071	0.067	0.064	0.062	0.062	0.051
Maximum	0.140	0.100	0.120	0.120	0.130	0.100	0.120	0.100	0.100	0.100	0.100	0.100	0.080

Heritage values were distributed throughout the weighting. Intangible heritage, customs and traditions are of great significance to Welsh culture and may explain the high weighting of historic value. Overlaps between historic and evidential value and authenticity can be hard to separate distinctly. This may have resulted in some clouding of which factors belong to the attribute in question, and thus the weighting. A definition of these values in the guidance for experts would be necessary for future use. The low evidential ranking could suggest that different forms of heritage values are considered more durable than physical fabric or perhaps the lighthouse's ruinous condition resulted in perceived lower evidential value, and any improvement would enhance its significance. Having further feedback from experts on the weighting results would be beneficial to understand their perspectives.

Scoring

Experts' qualitative values (Table 4.11) were transformed by value functions (Table 4.12) into quantitative values (Table 4.13). These were then multiplied by the mean weight of the respective attribute, as given by the equation:

$$\mathbf{V(a) = \Sigma w_i \cdot v_i(a_i)}$$

The results of the scoring demonstrated that adaptive release resulted in the highest score: 0.6236. Minimal intervention (0.6076) and restoration (0.5965) were in second and third place. Relocation of any type was shown to be less favourable, and demolition scored unfavourably. These results indicate that the optimum intervention (alternative) would be adaptive/planned release (Table 4.14).

The sensitivity analysis of the scoring was run twice, initially (S_1) increasing each attribute in turn to 0.19, while keeping the others at 0.0675. The process was completed with each attribute prioritised in turn, reflecting the skew of results (Table 4.16). The test was conducted a second time (S_2), with the maximum weight at 0.4 and all other weights set at 0.05 (Table 4.17).

As shown in Figures 4.10 and 4.11, S_1 produced a clear demarcation between the top three alternatives and the others. S_2, however, with the more extreme range between high and low weighting, demonstrates closer relationships between the scores.

Table 4.16: Results of S_1 sensitivity test. Copyright: Bella Romain.

	Minimal Intervention	Restoration	Relocation within setting	Relocation to OAM	Planned Release	Demolition
Historic	0.551	0.611	0.480	0.445	0.564	0.214
Evidential	0.551	0.435	0.389	0.368	0.357	0.351
Aesthetic	0.551	0.652	0.522	0.487	0.564	0.214
Communal	0.551	0.652	0.522	0.528	0.605	0.214
Opportunity for Engagement	0.603	0.652	0.562	0.528	0.615	0.275
Accessibility	0.511	0.529	0.501	0.528	0.523	0.214
Short Term Costs	0.633	0.590	0.439	0.405	0.646	0.275
Long Term Costs	0.633	0.529	0.501	0.466	0.646	0.336
Funding Opportunities	0.511	0.652	0.562	0.466	0.584	0.214
Risks to HE/ Archaeology	0.633	0.652	0.522	0.445	0.605	0.296
Risks to Environment	0.633	0.652	0.439	0.487	0.605	0.214
Impact on Authenticity	0.572	0.529	0.501	0.405	0.584	0.214
Logistical Possibility	0.633	0.529	0.439	0.405	0.646	0.275
	0.582	0.590	0.491	0.459	0.580	0.254

Table 4.17: Results of S_2 sensitivity test. Copyright: Bella Romain.

	Minimal Intervention	Restoration	Relocation within setting	Relocation to OAM	Planned Release	Demolition
Historic	0.494	0.627	0.441	0.416	0.503	0.159
Evidential	0.494	0.627	0.560	0.300	0.503	0.159
Aesthetic	0.494	0.742	0.560	0.535	0.503	0.159
Communal	0.494	0.742	0.560	0.650	0.622	0.159
Opportunity for Engagement	0.641	0.742	0.675	0.650	0.650	0.334
Accessibility	0.379	0.392	0.501	0.650	0.388	0.159
Short Term Costs	0.729	0.567	0.326	0.300	0.738	0.334
Long Term Costs	0.729	0.392	0.501	0.475	0.738	0.509
Funding Opportunities	0.379	0.742	0.676	0.475	0.563	0.159
Risks to HE/ Archaeology	0.729	0.742	0.560	0.416	0.622	0.393
Risks to Environment	0.729	0.742	0.326	0.535	0.622	0.159
Impact on Authenticity	0.554	0.392	0.501	0.300	0.563	0.159
Logistical Possibility	0.729	0.392	0.326	0.300	0.738	0.334
	0.582	0.603	0.501	0.462	0.596	0.244

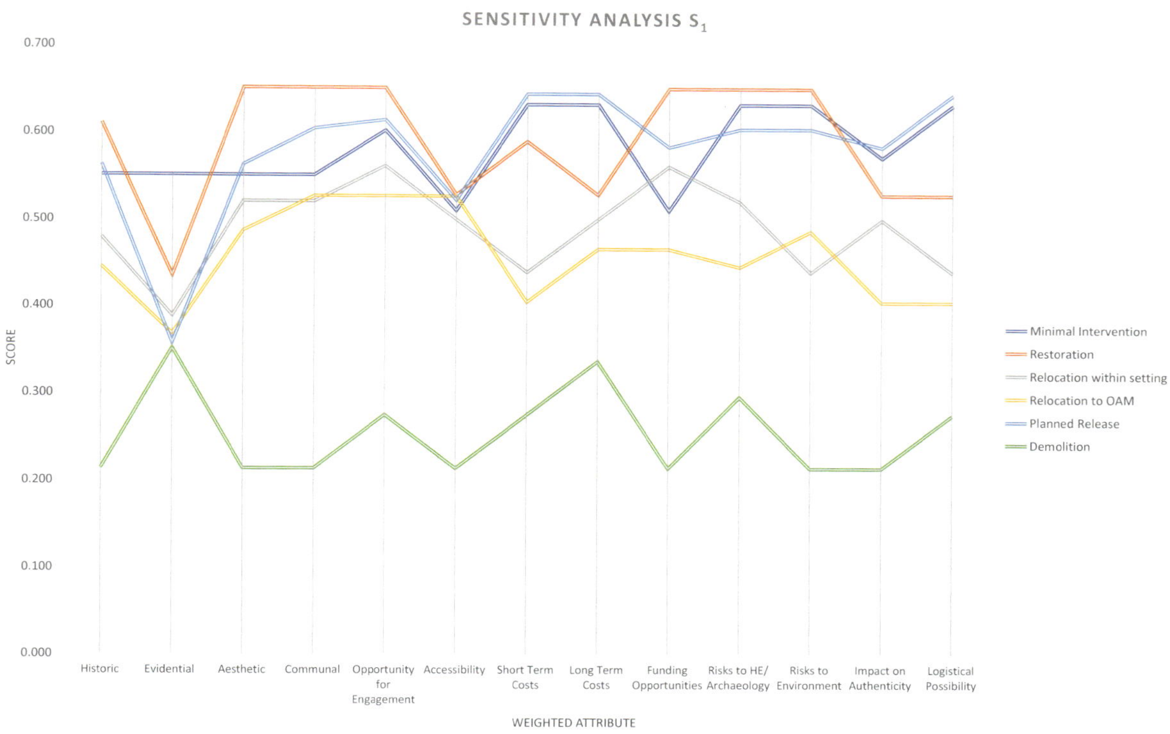

Figure 4.10: Graph showing S_1 sensitivity analysis results. Copyright: Bella Romain.

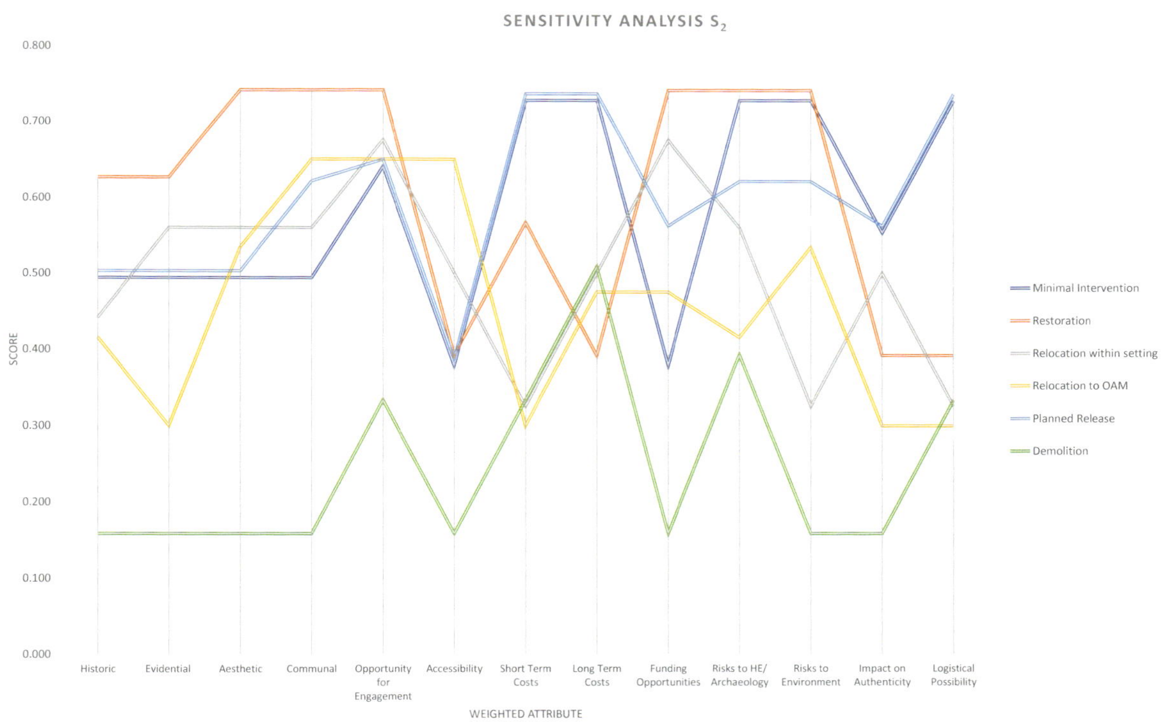

Figure 4.11: Graph showing S_2 sensitivity analysis results. Copyright: Bella Romain.

Table 4.18 indicates that demolition and relocation were the least preferred alternatives. The remaining alternatives scored highest in every version of the sensitivity analysis. Although sensitivity testing indicates restoration and minimal intervention are contenders to the MAVT-selected alternative, namely adaptive/planned release, when a mean is calculated from the results, it shows that the scores for that option are consistently high enough to skew the data in its favour. In more detail, the results were analysed to determine the alternatives' overall rank score, namely the cumulative total of all their rank positions across all scenarios, whereby the lower the score, the better the alternative. Results from S_1 (Table 4.18) indicate a clear preference for restoration, followed narrowly by planned release and minimal intervention (Figure 4.12).

Table 4.18: S_1 sensitivity ranking. Copyright: Bella Romain.

Sensitivity Ranking						
Sensitivity Test ↓	Minimal Intervention	Restoration	Relocation within setting	Relocation to OAM	Planned Release	Demolition
1	3	1	4	5	2	6
2	1	2	3	4	5	6
3	3	1	4	5	2	6
4	3	1	5	4	2	6
5	3	1	4	5	2	6
6	4	1	5	2	3	6
7	2	3	4	5	1	6
8	2	3	4	5	1	6
9	4	1	3	5	2	6
10	2	1	4	5	3	6
11	2	3	4	5	1	6
12	2	3	4	5	1	6
13	2	1	5	4	3	6
Totals	**33**	**22**	**53**	**59**	**28**	**78**

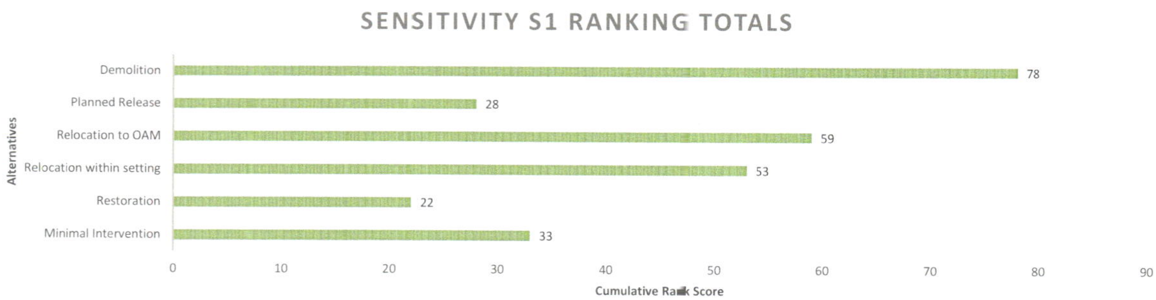

Figure 4.12: S_1 sensitivity ranking chart. Copyright: Bella Romain.

Results from S_2 (Table 4.19) show a narrower banding of aggregated scores. Demolition is still the least preferred alternative, but the other results are clustered. Restoration and planned release remain in Rank 1 (R1) and 2 (R2), but there is a shift in the scoring of minimal intervention and relocation within setting, and the scores are much tighter (Figure 4.13).

Table 4.19: S_2 sensitivity ranking. Copyright: Bella Romain.

Sensitivity Ranking						
Sensitivity Test ↓	Minimal Intervention	Restoration	Relocation within setting	Relocation to OAM	Planned Release	Demolition
1	3	1	4	5	2	6
2	4	1	2	5	3	6
3	5	1	2	3	4	6
4	5	1	4	2	3	6
5	5	1	2	4	3	6
6	5	3	2	1	4	6
7	2	3	5	6	1	4
8	2	6	4	5	1	3
9	5	1	2	4	3	6
10	2	1	4	5	3	6
11	2	1	5	4	3	6
12	2	4	3	5	1	6
13	3	1	4	5	2	6
Totals	**45**	**25**	**43**	**54**	**33**	**73**

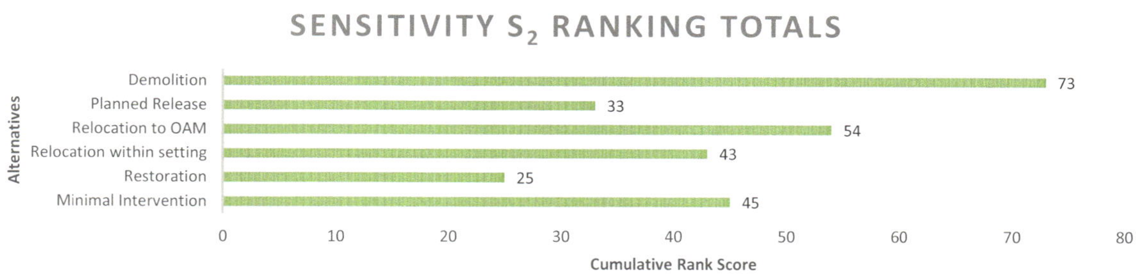

Figure 4.13: S_2 sensitivity ranking chart. Copyright: Bella Romain.

The rank-frequency, namely how many times an alternative ranked in a particular position, for S_1 (Table 4.20 and Figure 4.14), demonstrates that restoration is most frequently scored as the preferred alternative.

Table 4.20: S_1 sensitivity frequency. Copyright: Bella Romain.

Frequency Chart						
Rank ↓	Minimal Intervention	Restoration	Relocation within setting	Relocation to OAM	Planned Release	Demolition
1	1	**8**	0	0	4	0
2	6	1	0	1	**5**	0
3	**4**	4	2	0	3	0
4	2	0	**8**	3	0	0
5	0	0	3	**9**	1	0
6	0	0	0	0	0	**13**

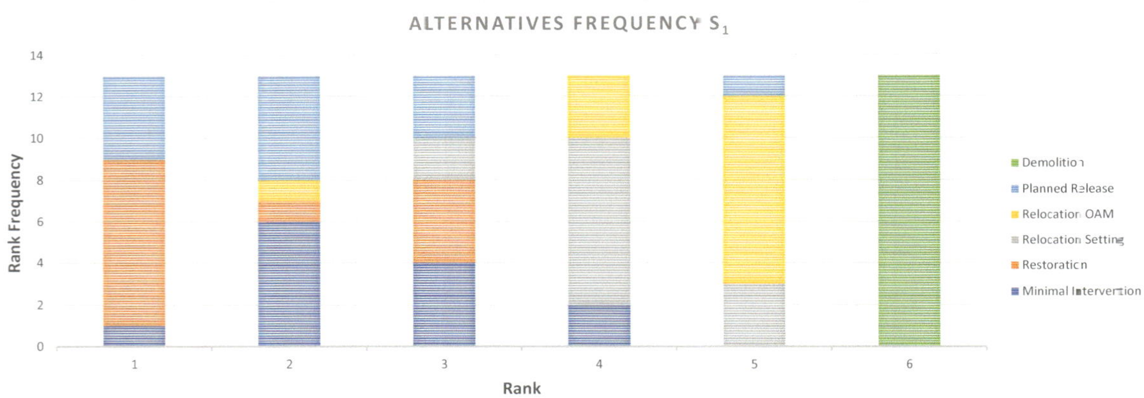

Figure 4.14: S_1 sensitivity frequency chart. Copyright: Bella Romain.

Minimal intervention and planned release have a narrow banding, but planned release is marginally ahead with a score of 9 in the top two ranks (R1 and R2), compared to minimal intervention with seven. Other placings are clear in the rank-frequency scores. Rank-frequency scores for S_2 (Table 4.21 and Figure 4.15) are less clear, with a wider distribution of placements. Restoration retains R1 overall, and although demolition is still the least preferable option (R6), it reached R3 in the weighting favouring long-term costs. Relocation to open air museum is R5. Although minimal intervention has five occurrences at R2, it is placed R4 overall due to five occurrences at R5. Similarly, relocation within setting has a wider distribution of scores, but the aggregated scores of higher ranks suggest that this is preferrable to minimal intervention. Conversely, although planned release only achieves two occurrences at R2, it achieves six at R3 and three at R1 and is therefore placed at R2 overall.

Table 4.21: S_2 sensitivity frequency. Copyright: Bella Romain.

Frequency Chart						
Rank ↓	Minimal Intervention	Restoration	Relocation within setting	Relocation to OAM	Planned Release	Demolition
1	0	**9**	0	1	3	0
2	**5**	0	**5**	1	2	0
3	1	2	1	2	**6**	1
4	2	1	**5**	2	2	1
5	**5**	0	2	**6**	0	0
6	0	1	0	1	0	**11**

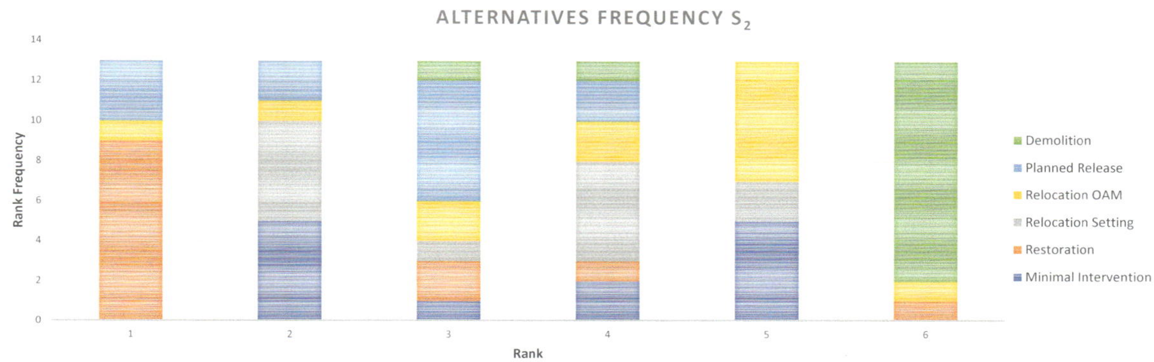

Figure 4.15: S_2 sensitivity frequency chart. Copyright: Bella Romain.

Comparison of the ranking frequencies (Table 4.22) demonstrates a slight preference for restoration overall (aggregated score of 5, with two R1 scores and 1 R3 score), despite planned release scoring highest in the MAVT (aggregated score of 5, with one R1 score and two R2 scores).

Table 4.22: Rank comparisons for S_1 and S_2. Copyright: Bella Romain.

Score	Minimal Intervention	Restoration	Relocation within setting	Relocation to OAM	Planned Release	Demolition
MAVT	2	3	4	5	1	6
S1	3	1	4	5	2	6
S2	3	1	4	5	2	6

Impact

The Whitford Point Lighthouse is a visible landmark across the estuary. Figure 4.5 shows a zone of theoretical visibility map, which accounts for landscape features and buildings in the area, showing that even with undulations in topography, it is visible from a considerable range. Whatever the eventuality for the conservation and management of the Whitford Point Lighthouse, the impact to the local community and landscape will be substantial.

Limitations

The MAVT analysis used to determine the preferred approach to conservation of the Whitford Point Lighthouse did not include any community engagement, which would have provided a more thorough perspective of the values and significance ascribed to it by all stakeholders. Another significant limitation was that not all identified and invited experts returned their MAVT assessments. Moreover, further discussion of the contributing factors and overall process with stakeholders would have been beneficial to ensure that everyone approached weighting with the same understanding, and to make clear the implications of each alternative. Further collaboration on the results of the MAVT analysis and its contribution would also have been advantageous to understand better the benefits of its wider use as a decision-making tool in heritage management and conservation.

The MAVT analysis in this research was tested on one heritage asset. The use of the methodology on multiple assets in the future, particularly where decisions have already been made, would offer insight into its efficacy on a wider scale.

Finally, through this research, it was realised that where an asset is particularly significant, but faces prohibitive challenges, the MAVT tool may not be able to determine clearly between extreme options, as demonstrated in the results with restoration and adaptive release favoured by most scenarios. This reinforces the use of MAVT as a supporting mechanism that guides decision making as opposed to a definitive analysis. Results such as these can inform decisions. For example, in the case of the Whitford Point Lighthouse, if there was a significant amount of funding available and economic viability was not a consideration, then restoration might be feasible. In the absence of these benefits, adaptive release is a justifiable alternative.

Implications

Given the lack of frameworks available for heritage managers to use in transferring adaptation policies to practice in the management of coastal heritage at risk, the MAVT tool and methodology used in this project could be a significant basis for identifying case studies for adaptive release. If this MAVT process works as well in practice as it does in theory, it could provide heritage managers with a decision-making support tool for complex heritage at risk cases. The tool delivers an objectively considered and justifiable alternative for the conservation management of complex assets and aligns closely with Cadw's Conservation Principles, which guides all conservation management in Wales.

Conclusion

Decisions on the future of built heritage at risk of inevitable loss can be supported using an adapted MAVT tool to score the available alternatives against determined attributes. The Whitford Point Lighthouse has been challenging to assess, but very appropriate for testing the potential of the MAVT methodology. As this study has shown, where decisions for conservation and management of heritage assets are straightforward, there is no need for MAVT analysis. Maintenance and conservation should always be the preferred approaches for heritage assets not at risk (DeSilvey 2023). MAVT for decision-making support appears to provide a robust analysis of the details and factors pertinent to individual cases. It is a simple tool which can be adapted in every way to reflect the specifics of heritage assets, its stakeholders, local community and landscape. This makes it flexible and applicable to a wide range of scenarios and potential alternatives. The compilation of information which is available to most heritage managers has the additional benefit of not being arduously time consuming or resource heavy. The most intensive aspect of the tool is the identification and involvement of stakeholders. Additional uses of the MAVT process that could benefit decision makers' processes include the allocation of funding to projects, determining the most appropriate asset for reuse, or prioritising maintenance works on assets at risk.

This research has proven that the MAVT methodology has a place in decision making for managing complex heritage assets. Further research should seek to establish its reliability and applicability to a range of subjects, in diverse environments and conditions. Although not intended to be used to identify heritage assets suitable for adaptive release, this might be a useful pathway and one which would ascertain its use as a decision-making tool for a range of heritage assets, particularly in light of the forthcoming challenges to be faced in the constructive loss of coastal heritage due to climate change.

'No strategy is ever innocent; none is inherently preferable over any other...Destruction and reconstruction are equally about the past-that-was and the future-to-be. Both are about values and meanings for people, about the end of past certainties and the uncertainties of new beginnings' (Holtorf 2018: 6).

References

Acanthus Holden 2008. Whitford Lighthouse Options Appraisal Report.

Aument, L. and J. Wathne 2012. Cast iron plate lighthouses and the Sakonnet Lighthouse restoration. *APT Bulletin: The Journal of Preservation Technology* 43.2-3: 3-11.

Avrami, E. 2009. Heritage, values, and sustainability, in A. Richmond and A. Bracker (eds) *Conservation: Principles, Dilemmas and Uncomfortable Truths*: 177-183. Abingdon: Routledge.

Avrami, E. 2020. Creative Destruction and the Social (Re) Construction of Heritage. *International Journal of Cultural Property* 27: 215-237. <https://doi.org/10.1017/S0940739120000120>

Barber, M. and Cawley, L., 2020, BBC News, Orfordness Lighthouse: Waving goodbye to a coastal landmark, viewed 16 August 2023, <https://www.bbc.co.uk/news/uk-england-suffolk-53263525>.

Bayley, J., 2017, Plymouth Live, Why Smeaton's Tower in Plymouth was moved to the Hoe, viewed 19 Augsut 2023, <https://www.plymouthherald.co.uk/news/smeatons-tower-plymouth-moved-hoe-633196>.

Cadw 2011. *Conservation Principles.* Cardiff: Cadw.

Cadw 2020. *Managing the marine historic environment of Wales.* Cardiff: Cadw.

Cadw 2023. Heritage at Risk data, email communication 28 October 2023.

Carmarthenshire Archive Service, 2023, Correspondence re Whitford Point. CAS/HT7.

Casey, A. and A. Becker 2019. Institutional and Conceptual Barriers to Climate Change Adaptation for Coastal Cultural Heritage. *Coastal Management* 47.2: 169-188. <https://doi.org/10.1080/08920753.2019.1564952>

Cassar, M. and R. Pender 2005. The impact of climate change on cultural heritage: Evidence and response. *14th Triennial Meeting: The Hague Preprints II, ICOM Committee for Conservation*: 610-616.

Chitty, G. 1987. A prospect of ruins. *Asociation for Studies in the Conservation of Historic Buildings* 12: 43-59.

Coflein, 2009, Swansea Harbour Lights West Pier, viewed 3 July 2023, <https://coflein.gov.uk/en/site/308198/?term=cast%2Biron%2Blighthouse>.

Colliers International 2011. Encouraging investment in industrial heritage at risk. Summary report 1 of 3, produced for English Heritage.

Cook, I. 2019. Climate change and cultural heritage: Developing a landscape scale vulnerability framework to measure and manage the impact of climate change on coastal historic landscapes. Unpublished PhD dissertation, University of Sheffield.

Cooke, C. 2000. What is the point of saving old buildings? *Arq. Architectural Research Quarterly* 4.2: 137-148.

Cunningham, T., 2013, Dublin Institute of Technology, Factors affecting the cost of building work- an overview, viewed 16 September 2023, <https://arrow.tudublin.ie/beschreoth/27/>.

Davidson, A. 2002. *The coastal archaeology of Wales*. York: Council for British Archaeology.

Dawson, T. 2016. Community Rescue: Saving Sites from the Sea. *Arquelogia Publica P: Online Journal in Public Archaeology* 2: 5-42. <https://doi.org/10.23914/ap.v6i2 78>

Dawson, T., J. Hambly, W. Lees and S. Miller 2021. Proposed policy guidelines for managing heritage at risk based on public engagement and communicating climate change. *The Historic Environment: Policy and Practice* 12. 3-4: 375-394. <https://doi.org/10.1080/17567505.2021.1963573>

De Blieu, J. and M. Halminski 2001. The taking of the Hatteras Light. *Southern Cultures* 7.3: 8-26. <https://dx.doi.org/10.1353/scu.2001.0035>

DeSilvey, C. 2006. Observed decay: Telling stories with mutable things. *Journal of Material Culture* 11.3. <https://doi.org/10.1177/1359183506068808>

DeSilvey, C., 2023, personal communication on practicalities of AR. Via MS Teams, with B. Romain, 26 July 2023. Interview not transcribed for confidentiality.

DeSilvey, C., H. Fredheim, H. Fluck, R. Hails, R. Harrison, I. Samuel and A. Blundell 2021. When loss is more: from managed decline to adaptive release. *The Historic Environment: Policy and Practice* 12.3-4: 418-433. <https://doi.org/10.1080/17567505.2021.1957263>

DeSilvey, C., H. Fredheim, A. Blundell and R. Harrison 2022. Identifying opportunities for integrated adaptive management of heritage change and transformation in England: a review of relevant policy and current practice. Historic England Research Report Series, 18/2022.

DeSilvey, C. and R. Harrison 2019. Anticipating loss: rethinking endangerment in heritage futures. *International Journal of Heritage Studies* 26.1: 1-7. <https://doi.org/10.1080/13527258.2019.1644530>

Deutsche Stiftung Denkmalschutz, 2023, Dauerhafte Bewahrung des Leuchtturm Roter Sand [Permanent Preservation of Roter Sand Lighthouse], viewed 31 July 2023, <https://www.denkmalschutz.de/fileadmin/media/Bilder/Denkmal/Leuchtturm_Roter_Sand/Praesentation_Bewahrung_Mai2023.pdf>.

Drozd, W. 2019. Structure relocation. *Civil and Environmental Engineering Reports* 29.4: 176-184. <https://doi.org/10.2478/ceer-2019-0053>

Earl, J. 2003. *Building Conservation Philosophy*, 3rd Edition. Shaftesbury: Donhead.

Edwards, N. 2023. Email communication to B. Romain re NRW licencing at Whitford Point Lighthouse, 9 August 2023.

Emerick, K. 2014. *Conserving and managing ancient monuments: heritage, democracy, and inclusion.* Woodbridge: The Boydell Press.

Emmerson, N. and D. Watkinson, 2022, Interview for MSc Assignment, interviewed by Faulkner, E., Conservation Plan: Whiteford Point Lighthouse, Gower Peninsula, South Wales: Heritage at Risk. Unpublished MSc Assignment, Cardiff University.

Eshelman, R. 1997. Lighthouse construction types, in R. M. Greenberg (ed.) *Keeping Lighthouses: A new breed of keepers focus on preservation, Cultural Resource Management* 20.8: 5-8. Washington: National Park Service.

Fatorić, S. and E. Seekamp 2017. Are cultural heritage and resources threatened by climate change? A systematic literature review. *Climatic Change* 142: 227-254.

Faulkner, E. 2022. Conservation Plan: Whiteford Point Lighthouse, Gower Peninsula, South Wales: Heritage at Risk. Unpublished Master's dissertation, Cardiff University.

Ferretti, V., M. Bottero and G. Mondini 2014. Decision making and cultural heritage: An application of the Multi-Attribute Value Theory for the reuse of historical buildings. Journal of Cultural Heritage 15.6: 644-655. <https://doi.org/10.1016/j.culher.2013.12.007>

Fishburn, P.C. 1967. *Additive Utilities with Incomplete Product Set: Applications to Priorities and Assignments.* Baltimore: Operations Research Society of America (ORSA) Publication.

Fluck, H. 2016. Climate Change Adaptation Plan. *Research Report Series*, no. 28/2016, Historic England.

Forster, A. M. and B. Kayan 2009. Maintenance for historic buildings: a current perspective. *Structural Survey* 27.3: 210-229.

Goblet, N. 2006. Moving historic buildings: One means of preservation. Unpublished Masters dissertation, Cornell University.

Gregory, D., T. Dawson, D. Elkin, H. Van Tilburgh, C. Underwood, V. Richards, A. Viduka, K. Westley, J. Wright and J. Hollesen 2022). Of time and tide: The complex impacts of climate change on coastal and underwater heritage. *Antiquity* 96.390: 1396-1411. <https://doi.org/10.15184/aqy.2022.115>

Gregory, J. 2008. Reconsidering relocated buildings: ICOMOS, authenticity and mass relocation. *International Journal of Heritage Studies* 14.2: 112-130.

Guest, K., I. Samuel and C. DeSilvey, 2022, Landscape Futures and the Challenge of Change, Report and Recommendations, viewed 12 November 2025, <https://www.exeter.ac.uk/v8media/research/esi/documents/Landscape_Futures_Report_and_Recommendations_August2022.pdf>.

Hague, D. 1972. Whitford Point Lighthouse. *Gower* 23: 44-46.

Hartley, J. 2003. The first Whitford Lighthouse. *Gower* 54: 16-32.

Hartley, J. 2005. The second Whitford Lighthouse. *Gower* 56: 13-34.

Hassan, M. and H. Xie 2020. Climate Change and Conservation of Coastal Built Heritage, in H. Erlich, M. Tayob, H. Mubarak, and S. A. Guerra (eds) *Climate Change and Conservation of Coastal Built Heritage*: 3-40. Singapore: Springer Singapore.

Hayles, C. S. 2022. *How resilient are buildings in the UK and Wales to the challenges associated with a changing climate?* Cardiff: Welsh Government.

Heesom, D., P. Boden, A. Hatfield, A. De Los Santos Melo, and F. Czarska-Chukwuran 2021. Implementing a HBIM approach to manage the translocation of heritage buildings. *Engineering, Construction and Architectural Management* 28.10: 2948-2966. <https://doi.org/10.1108/ECAM-06-2020-0405>

Hellier, N. 2023. Personal communication on options for conservation of Whitford Point Lighthouse. Via MS Teams, with B. Romain, 7 July 2023. Interview not transcribed for confidentiality.

Historic England, 2013, Heritage at risk, viewed 6 June 2023, <https://historicengland.org.uk/advice/heritage-at-risk/buildings-at-risk/>.

Historic Environment Group, Climate Change Subgroup, 2020, Historic Environment and Climate Change in Wales, Sector Adaptation Plan (WG39783), viewed 11 June 2023, <https://cadw.gov.wales/advice-support/climate-change/adapting-to-climate-change>.

Holtorf, C. 2018. Conservation and heritage as future making, in C. Holtorf, L. Kealy, and T. Kono (eds) *Conservation and heritage as future making, ICOMOS University Forum*: 1-13.

Howe, G. 1997. Report on Whitford Point. WGAS D/D_Z_850

Hunt, R., 2023, The Blackpool Gazette, Artist hoping to prevent ruined Fleetwood lighthouse Wyre Light being lost forever, viewed 7 May 2023, <https://www.blackpoolgazette.co.uk/news/people/artist-hoping-to-prevent-ruined-fleetwood-lighthouse-wyre-light-being-lost-forever-4129435>.

ICCROM 2000. The Riga Charter.

ICOMOS 1931. The Athens Charter for the Restoration of Historic Monuments.

ICOMOS 1994. The Nara Document on Authenticity.

ICOMOS 2013. The Burra Charter: The Australia ICOMOS Charter for Places of Cultural Significance.

ICOMOS 2020. *Heritage at risk: World report 2016-2019 on monuments and sites in danger*. Berlin: hendrik Bäßler verlag.

ICOMOS and TICCIH 2011. The Dublin Principles, Joint ICOMOS – TICCIH Principles for the Conservation of Industrial Heritage Sites, Structures, Areas and Landscapes.

Jackier, D.A. 2001. Moving Historic Structures: The Washington Street Renewal Project. Unpublished Masters dissertation, University of Pennsylvania.

Kaltenbacher, T. 1997. Pooles Island Lighthouse Stabilisation Project, in R. M. Greenberg (ed.) *Keeping Lighthouses: A new breed of keepers focus on preservation, Cultural Resource Management* 20.8: 12-14. Washington: National Park Service.

Keeney, R. L. and H. Raiffa 1976. *Decisions with multiple objectives: Preferences and value tradeoffs*. Cambridge: Cambridge University Press.

Kerr, J.S. 2013. The Seventh Edition Conservation Plan, ICOMOS.

Khakban, M., B. Pedram and M. A. Emami 2020. The challenge of the authenticity of the work and the environment around it in open air museums. *Bagh-e Nazar* 17.89: 49-62.

Khalaf, R. W. 2022. Periodic reporting under the World Heritage Convention: Futures and possible responses to loss. *The Historic Environment: Policy and Practice* 13.3: 341-360.

Lewis, M. 2012. Iron Lighthouses. *Construction History* 23: 23-64.

Llanelly Harbour Act 1896, viewed 12 October 2023, <https://www.legislation.gov.uk/ukla/Vict/59-60/186/contents/enacted>.

Loader, R.D. 2011 Quantifying coastal heritage change. Unpublished PhD dissertation, University of Exeter.

Lowe, J.A. and J. M. Gregory 2005. The effects of climate change on storm surges around the United Kingdom. *Philosophical Transactions: Mathematical Physical and Engineering Sciences* 363.1831: 1313-1328. <https://doi.org/10.1098/rsta.2005.1570>

Lowe, J.A., J. M. Gregory, and R. A. Flather 2001. Changes in the occurrences of storm surges around the United Kingdom under a future climate scenario using a dynamic storm surge model driven by the Hadley Centre climate models. *Climate Dynamics* 18: 179-188.

MacDonald, S.M. 2018. Sentinels by the sea: keeping as an alternative tourism performance. *Text and Performance Quarterly* 38.1-2: 19-37.

Matero, F.G. 1993. The Conservation of Immovable Cultural Property: Ethical and Practical Dilemmas. *Journal of the American Institute for Conservation* 32.1: 15-21.

Morris, W., Webb, P., Society for the Protection of Ancient Buildings, Manifesto, viewed 17 October 2023 <https://www.spab.org.uk/about-us/spab-manifesto>.

Muños Viñas, S. 2009. Minimal intervention revisited, in Richmond, A. and A. Bracker (eds) *Conservation: Principles, Dilemmas and Uncomfortable Truths*: 47-59. Abingdon: Routledge.

Murphy, P., D. Thackray and E. Wilson 2009. Coastal heritage and climate change in England: Assessing threats and priorities. *Conservation and Management of Archaeological Sites* 11.1: 9-15.

National Trust 2007. *Shifting shores: Living with a changing coastline*. Cardiff: The National Trust Wales.

Nevell, M. 2010. Industrial heritage at risk. *Industrial Archaeology Review* 33.2: 79-80.

Nicholas, F. 1997. Report on a survey of Whiteford Point Lighthouse, Gower. Gower Preservation Trust.

Oil and Gas UK, 2013, The Management of Marine Growth During Decommissioning. Oil and Gas UK.

Pembrokeshire Coast National Park, Changing coasts, Pembrokeshire Coast National Park, viewed 7 August 2023, <https://www.pembrokeshirecoast.wales/get-involved/changing-coasts/>.

Peltola, X. 2008. Moving historic buildings: A study of what makes good preservation practices when dealing with historically significant buildings and structures. Unpublished Masters dissertation, Clemson University.

Piazzoni, F. 2020. What's Wrong with Fakes? Heritage Reconstructions, Authenticity, and Democracy in Post-Disaster Recoveries. *International Journal of Cultural Property* 27: 239-258. <https://doi.org/10.1017/S0940739120000119>

Romain, B. 2023. Against the Dying of the Light: How do we make decisions on the future of built heritage which is at risk of inevitable loss? Using Whitford Point Lighthouse to develop a decision-making tool for Heritage at Risk. Unpublished Master's dissertation, University of York.

Smith, M. 2014. Citizen science in archaeology. *American Antiquity* 79.4: 749-762.

Smith, S., 2023. Personal communication on options for conservation of Whitford Point Lighthouse. Via MS Teams, with B. Romain, 7 July 2023. Interview not transcribed for confidentiality.

Stanley-Price, N. 2009. The Reconstruction of Ruins: Principles and Practice, in A. Richmond and A. Bracker (eds) *Conservation: Principles, Dilemmas and Uncomfortable Truths*: 32-46. Abingdon: Routledge.

Storm, A. 2008. Hope and rust: Reinventing the industrial place in the late 20th century. *Stockholm Papers in the History and Philosophy of Technology*, TRITA-HOT-2057.

Swansea Council, 2013, Listed Buildings at Risk in the City and County of Swansea, viewed 12 August 2013, <https://www.swansea.gov.uk/article/9600/Listed-buildings-and-ancient-monuments-at-risk>.

Swansea Council, 2016, Listed building at risk register, viewed 12 August 2023, <https://www.swansea.gov.uk/media/2829/Listed-building-at-risk-register-May-2016/pdf/Listed_building_at_risk_register_May_2016.pdf?m=1630593376317>.

The Llanelli Harbour Revision Order 1997, (SI 1997/325), viewed 16 October 2023, <https://www.legislation.gov.uk/uksi/1997/325/made>.

Van Dyke, S. 2004. The history of wrought and cast iron. Unpublished Master's dissertation, University of Tennessee.

Venture, T., C. DeSilvey, B. Onciul and H. Fluck 2021. Articulating loss - A thematic framework for understanding coastal heritage transformations. *The Historic Environment: Policy and Practice* 12.3-4: 395-417. <https://doi.org/10.1080/17567505.2021.1944567>

Walker, D. 2011. Towards a beneficial World Heritage: community involvement in the Blaenavon industrial landscape. *Museum International* 63.1: 25-33. <https://doi.org/10.1111/j.1468-0033.2012.01760.x>

Wallace, R.D. 2009. Lighthouse memories in Texas: Selective memories, selective preservation. Unpublished Master's dissertation, Texas Tech University.

Welsh Coastal Monitoring Centre, Coast Snap, viewed 7 August 2023, <https://www.wcmc.wales/coastsnap>.

Welsh Government, 2023, FOI release - Listed buildings, viewed 13 January 2025, <https://www.gov.wales/atisn19140>.

Welsh Government, 2024, FOI release - Buildings at risk, viewed 13 January 2025, <https://www.gov.wales/atisn19270>.

Whitewright, J. 2023. Personal communication on options for conservation of Whitford Point Lighthouse. Via MS Teams, with B. Romain, 7 July 2023. Interview not transcribed for confidentiality.

Wienberg, J. 2014. Four churches and a lighthouse. *Danish Journal of Archaeology* 3.1: 68-75. <https://doi.org/10.1080/21662282.2013.910366>

5.
Transdisciplinary Approaches to Understanding Climate Change Impacts on Heritage in the Colombian Caribbean: The Case of Cartagena de Indias

L.Victoria Báez Santos, Daniela Acosta Romero, Jesús Aldana Mendoza, Carlos Del Cairo Hurtado, Carla Riera Andreu

Cartagena de Indias, in Colombia, has increasingly faced severe weather events in recent years, causing significant damage and partial loss of its maritime cultural heritage. The project *Colaborartorio Azul* is an initiative by the NGO *Fundación Colombia Anfibia* that seeks to create a comprehensive methodology that integrates archaeological, historical, anthropological, geographical, geological, oceanographic, and hydrographic data with local knowledge and expertise to assess the effects of climate change on the city's tangible and intangible heritage. By utilising diverse data sources, including historical cartography, aerial photographs, oceanographic reports, and contemporary local experiences, this study discusses coastline transformations, the physical impacts of climatic events, cultural resource degradation, community adaptations, and mitigation strategies through a diachronic perspective. Since the project is still in progress, this paper focuses on the team's transdisciplinary approaches emphasising the theoretical and methodological frameworks used with the aim to provide a holistic identification and documentation of climate change impacts on the heritage of Cartagena de Indias.

Introduction

In recent years, scientists have been documenting the severe impacts of climate change on land, riverine, marine, and coastal environments. Cultural heritage preserved in these environments has been damaged from warming temperatures and abnormal weather patterns, making climate change the fastest growing threat to heritage assets globally (Kimmell and Markham 2021). Given the risk that climate change poses to heritage, several studies have highlighted its vulnerability to various climate change effects, demonstrating the need for mitigation and adaptation plans (e.g., Barba Meinecke *et al.* 2010; Ezcurra and Rivera 2018; ICOMOS 2020; Reeder-Myers 2015; UNESCO 2022).

In Latin America and the Caribbean, several initiatives seek to create successful strategies for research, management and sustainability of heritage assets in the face of the climate emergency. For example, in Cuba, there has been a proposal to create climate change indicators that can measure and map the impacts of climate change on maritime heritage along with risks posed by tourism pressures, real estate development or other changes in the local environment and communities (Brito *et al.* 2023). Similarly, the DUNAS (Descendants United for Nature, Adaptation, and Sustainability) project, based in Puerto Rico, seeks to initiate community resilience by involving local populations, firstly, in the restoration of local natural ecosystems at the coast, such as sand dunes, and, secondly, in the study of cultural heritage, critical to community identity, and the risks posed by climate change effects (Climate Science Alliance – DUNAS 2025). Hence, the past is perceived and utilised as a tool that can decolonise historical narratives, while also answering local questions and addressing climate-related problems for the present and the future (Rivera-Collazo 2022).

Cartagena de Indias (Figure 5.1), the case study of this project in Colombia, is a city with an immense wealth of cultural heritage, both tangible and intangible. It is a UNESCO World Heritage Site since 1984, and part of the 'Port, Fortresses and Group of Monuments', adhering to selection criteria IV and VI (UNESCO, The Criteria for Selection n.d.). The city stands out as an 'example of the military architecture of the 16th, 17th, and 18th centuries, the most extensive of the New World and one of the most complete', as well as for its role as 'essential link in the route of the West Indies' (UNESCO n.d.).

Currently, Cartagena de Indias is severely affected by climate change. Coastal erosion, sea level rise, flooding, temperature increase, and the increase of extreme weather events (torrential rains, heatwaves, tidal waves) are just a few of the phenomena impacting the built environment and the lives of the local inhabitants. According to the city's 4C Plan ('A Competitive and Climate Compatible Cartagena'), it is predicted that by 2040 the sea level will rise by 15cm to 20cm, there will be 30% more heavy and frequent rainfall, and the average temperature of the city will reach 29.1°C as opposed to the 27.9°C observed in 2010. Under this scenario, in 2040, neighbourhoods like Ciénaga de La Virgen, Bocagrande, Castillo Grande, El Laguito, the historic centre, the port and industrial zones will be flooded and/or partially submerged, while areas such as the island of Tierrabomba, Playetas, Punta Gigante (Baru), Isla Fuerte, the Archipelago of Islas del Rosario and San Bernardo will suffer from major erosion (Alcaldía de Cartagena de Indias 2021). With these projections, it is estimated that 86% of the historical structures and heritage of the city is under threat (Alcaldía de Cartagena de Indias 2021). To take measures against climate change and its impacts on the cultural heritage of the city, government authorities incorporated within the 4C Plan the strategy for the 'Protection of Historical Heritage'. This strategy assesses how the historical heritage of Cartagena de Indias could become resilient to climate change effects by 2040. This would require actions framed in a climate compatible development that maintains the city's value as a World Heritage Site and Cultural Interest Asset at the service of Cartagena's residents and visitors.

The 'Blue Collaboratory' project (*Colaborartorio Azul* in Spanish), presented in this study, was created to align with the objectives of the above strategy, with the main aim to contribute to the preservation and protection of the maritime cultural heritage of Cartagena de Indias during the climate emergency (Colaborartorio Azul 2025; Del Cairo *et al.* 2024). This initiative covers both urban and rural areas, employing a comprehensive approach that involves detailed analysis and in-depth study of the climate change effects impacting tangible and intangible heritage. Furthermore, the project aims at developing a robust strategy for monitoring these changes in the long term, ensuring sustained safeguarding of the local invaluable heritage for future generations.

This paper introduces *Colaborartorio Azul* and its significance. After explaining its aims and methodology the chapter proceeds with more detailed observations regarding climate change impacts on the maritime cultural heritage of Cartagena de Indias. Subsequently, the importance of transdisciplinary methodologies is explained along with reflections on creative ways for integrating local and indigenous knowledge of the residents of Cartagena de Indias in cultural heritage management during the climate emergency.

Aims and methodology

Colaborartorio Azul is part of the Interadministrative Agreement (CI-IPCC-003-2022) between the Institute of Cultural Heritage of Cartagena and the University of Cartagena. Its main objective has been to identify and analyse the impacts of climate change on the cultural heritage of Tierrabomba island in Cartagena de Indias, through a transdisciplinary and interinstitutional approach (Del Cairo *et al.* 2023a,

2023b). The study area includes the towns of Tierrabomba, Bocachica, Caño del Oro, the historic centre of Cartagena De Indias, Castillo Grande, and El Laguito (Figure 5.1).

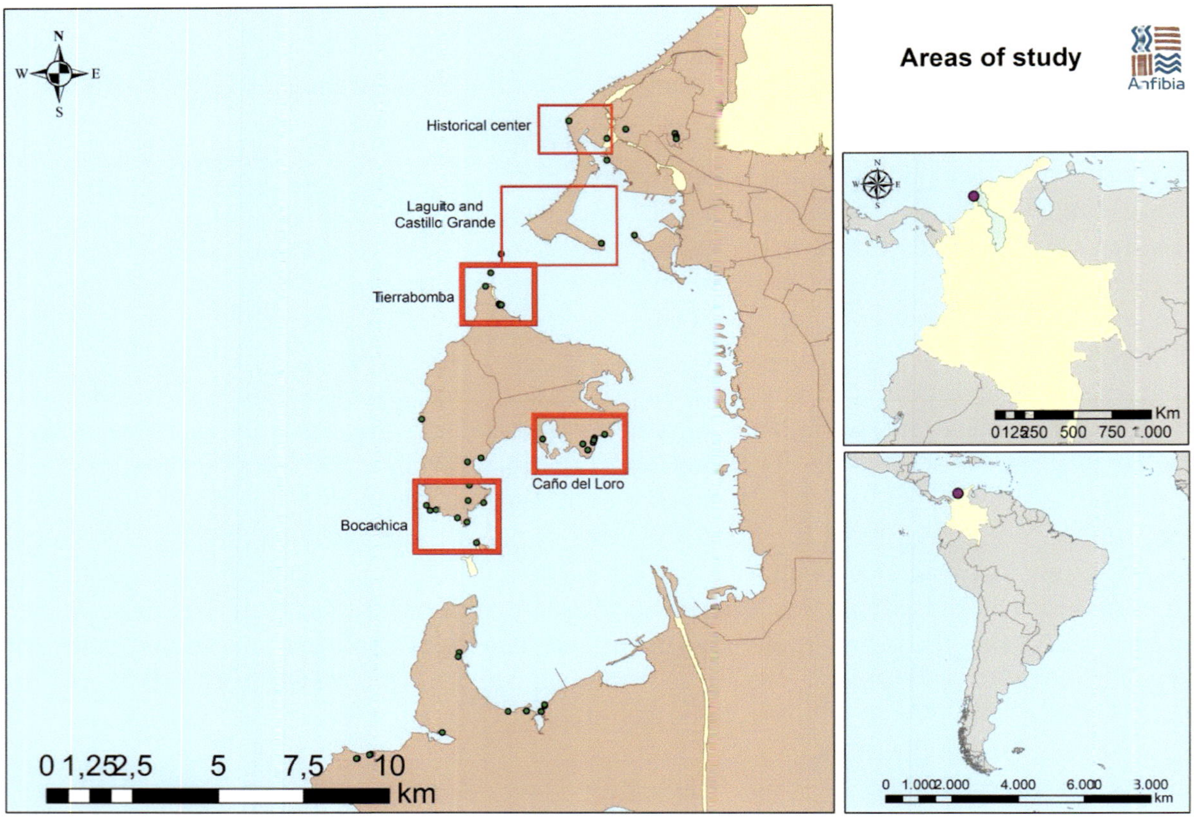

Figure 5.1: Study area of Colaborartorio Azul in Colombia. Copyright: L. Victoria Báez Santos.

Colaborartorio Azul adopts a participatory training component to recognising risks to heritage. This enables capacity building of local communities and dissemination of knowledge for the identification and monitoring of the impacts of climate change on the tangible and intangible maritime and underwater heritage of the region. Because of this, the project is aligned with the recommendations established by the Action Plan for World Heritage in Latin America and the Caribbean (2014-2024) (UNESCO 2014), which highlights the need to create studies for the identification of management indicators for the impacts of climate change and disaster management projects on cultural heritage through the strengthening of local capacities, development of lines of action and resilience to these phenomena. In addition, the project is contributing to Sustainable Development Goal 4 ('Quality Education') by building local capacity to identify, assess and monitor risks to heritage, in particular considering Target 7, which mentions the importance of acquiring theoretical and practical knowledge for 'appreciation of cultural diversity and of culture's contribution to sustainable development' (UN SDGs n.d.). It also addresses Sustainable Development Goal 11 ('Sustainable Cities and Communities') and its Target 4, which recommends to 'strengthen efforts to protect and safeguard the world's cultural and natural heritage'.

The project has two main phases. In the first phase, the various forms of knowledge from institutional representatives, community members of various neighbourhoods and age groups, as well as scientists were brought together through meetings and workshops to discuss the impacts of climate change on various heritage resources and to help identify observation stations around the city, where it would be possible to track such climate change impacts. The second phase aims at co-creating a proposal for successful evaluation and monitoring strategies to measure the vulnerability of heritage sites to climate change effects.

This paper focuses mainly on the methodological approaches of the project and the results of the first phase described above. With this work we hope to emphasise the importance of transdisciplinarity, community participation and co-creation in this type of initiatives, along with the challenges seen so far in our case.

Understanding climate change impacts on the maritime cultural landscape

Colaborartorio Azul addresses the effects of climate change on 'assets of cultural interest' that are part of the maritime cultural landscape[1] (as defined by Westerdahl 1992; 2011) of Cartagena de Indias. This landscape perspective offers the opportunity to consider in more detail the interlinks between land and sea, the relationship between heritage, communities, and the different environmental phenomena that occur there, as well as the current climate change. Previous studies (e.g., Cook 2019) have demonstrated how cultural landscapes are vulnerable to climate change due to direct and indirect factors that lead to the loss of tangible and intangible heritage, changes related to the practices of local communities affecting the lifestyles of the region's inhabitants, as well as changes in the visual characteristics of the landscape and local ecosystems. For this reason, the climate change management plan 4C in Cartagena (Alcaldía de Cartagena de Indias 2021) has been incorporated in *Colaborartorio Azul* since it considers assets of cultural interest along Cartagena's coastal strip under threat by climate change, including the historic centre, showing that the transformations that the region is undergoing can drastically affect heritage resources, mainly tangible, but also intangible, associated with the cultural capital of a large population linked to life by the sea for several centuries.

Configuring the relationship between heritage, communities and climate change in practice and within policy presents a big challenge, since local communities have different ways of appropriating their environment, both on land and at sea, where the economic, institutional and symbolic spheres come into play and are articulated. This is how 'a set of constructs configured from relationships with the sea and among themselves, which have a historical anchorage and are at the same time the object of change and evolution, are specified' (Rubio-Ardanáz 2014: 28). From that it becomes obvious that in recent decades, climate change has not only put cultural heritage under thread, but it has strained and shifted the relationship of communities with their surrounding natural environment.

To address the relationship between heritage, communities and climate change, *Colaborartorio Azul* follows transdisciplinary, participatory and co-creative approaches with the aim to generate knowledge about climate change and its impact on both natural and cultural heritage in the city of Cartagena de Indias. Transdisciplinarity provides a holistic approach to complex scenarios or issues by bridging the work conducted in different disciplines and crossing the boundaries of communication and collaboration between them. In *Colaborartorio Azul*, through the use of a variety of expertises and disciplines (including archaeology, anthropology, history, geography, climate science, oceanography, hydrography, conservation and ecology) and methodologies (such as mapping, ethnographical recording, oral histories, photographic documentation, water/sea level monitoring, and participatory workshops) our team has been able to take on the challenge of addressing climate change and its effects on maritime and underwater cultural heritage in Cartagena, and thus act jointly with the community

[1] This is a concept that understands maritime landscapes as a sphere in the interface between land and sea and as an element loaded with meanings, cultural representations, which is also constantly subject to transformations caused by human and non-human agents.

to make plans for safeguarding the tangible cultural resources, the associated intangible aspects and the natural environment. This follows Sánchez and Cardona's (2017: 50) observation that with transdisciplinary approaches 'there are spaces of contact and cooperation between various overlapping specialties to lead to a unified paradigm that manages to reconceptualise and reconcile classic dichotomies-such as the symbolic and the material, objectivity and subjectivity, theory and practice'.

By intertwining all the above elements, we generate data that help us understand the effects of climate change on heritage, taking into account the different dimensions that contribute to it and, above all, those who are present and inhabit the bay of Cartagena. Human actions and behaviours are central to this since they directly influence climate variability and levels of risk. Therefore, we consider it necessary to actively involve the community in our study and adopt a co-creative and participatory approach that allows the interweaving of local and indigenous knowledge, while accounting for new behaviours and other human factors that exist for the local populations, their heritage and the natural environment.

Figure 5.2: Workshop with local stakeholders using newspaper archives from the 20th century in 2023. Copyright: Gabriela Caro.

First attempts in identifying and monitoring climate change impacts on the maritime cultural heritage of Cartagena de Indias

Colaborartorio Azul developed different activities that covered the social, environmental, historical, and heritage aspects of Cartagena de Indias and its population. At first, desk-based research on the relevant actors, resources, and risks associated with cultural heritage in the territory of Tierrabomba island was carried out. For this, we followed the approach proposed by the International Centre for the Study of the Preservation and Restoration of Cultural Property (ICCROM) in 2015, which emphasized the conservation of cultural heritage through people-centred methodologies. Hence, the project developed a practical fieldwork element that aimed at involving inhabitants of the island of Tierrabomba (from the villages of Bocachica, Tierrabomba and Caño del Loro) in the research by forming working groups and other community interactions. Additionally, we partnered with the community foundation *Vigías de Karex* and the foundation *Los Jagüeyes* and reached out to personnel from the Maritime Authority (DIMAR - *Dirección General Marítima* in Spanish), the ECTAR (*Escuela Taller de Cartagena de Indias*) and the Caribbean Naval Museum, while also bringing in our academic partners from the University of Cartagena de Indias. Through a series of meetings (Figures 5.2-5.3), we initiated discussions and collected data on the experiences of the contemporary communities and experts on the impacts of climate change on the maritime cultural resources of the region.

Moreover, the project attempted to reconstruct natural vulnerabilities of the coast of Cartagena over a longer timescale. Through the analysis of primary historical sources (such as historical maps, archival documents and available archaeological records), it was possible to recognise natural characteristics of the region, along with climatic features, natural phenomena and their frequency, as well as natural and anthropogenic transformations of the terrain and the associated cultural resources over time. Through discussions with local residents (Figures 5.4-5.5) and evidence from archival material it was also possible to take into consideration the diverse environmental perceptions of Cartagena's inhabitants. Thus, it was possible, for example, to identify shoreline erosion and accretion in different study areas of Cartagena, showing major changes over time due to anthropogenic and natural factors. The threat of erosion in this territory has already been addressed by other authors (e.g., Afanador *et al.* 2008; Ricaurte *et al.* 2018), who highlight the difficulties caused by the few protective structures and the influence of activities, such as boat traffic, on this phenomenon.

Through our study it was also possible to define and characterize some of the most affected areas of the coast. This became possible through information collected from a variety of documentary sources, such as historical maps and aerial photographs that identify the accumulation of sediments and transformations of the territory in past and recent years, indicating a possible process of intensive coastal erosion since the beginning of the 20th century that has resulted in the partial or complete loss of some sections of the coastline (Del Cairo *et al.* 2023a; 2023b). For example, in areas such as the north of the island of Tierrabomba, although a region with cyclical transformations of the coastline, historical mapping allowed us to identify that in recent years (1979-2021) there has been a significant increase in the rate of erosion. This, complemented by the gradual increase in temperature in the city (with the 2015 and 2016 temperatures being the hottest recorded so far), puts the coast and all its assets at immense risk. With that in mind, we started identifying heritage assets and other significant material culture, such as forts, batteries, civil infrastructures, and isolated sites on the coastline or in intertidal spaces, that might be impacted in the course of the 21st century.

Figure 5.3: Workshop using the 'Mandala Methodology' in 2023. Copyright: Gabriela Caro.

Figure 5.4: Discussions and interviews with local residents. Copyright: Colaborartorio Azul.

Figure 5.5: Discussions and interviews with local residents. Copyright: Colaborartorio Azul.

Additionally, from the working groups formed with members of the local community, it was possible to document the environmental variables that have most affected people's lives and heritage in the area and their perception of these changes. Overall, the community seemed to recognise that the effects they are experiencing are due to atmospheric and marine phenomena such as storms and hurricanes, strong winds, floods, high tides. These have caused serious damage to archaeological sites and heritage, with observations such as damages to masonry, deterioration due to humidity, and colonization of fauna and flora on historical structures during periods of flooding. This information was complemented by a series of field visits with the community co-researchers (Figure 5.6) to the relevant archaeological and heritage sites, where in-person documentation was carried out. Thus, it was possible to observe occurrences such as the decomposition and collapse of some structures (for example, at the tilery de San Bernabé), the impact of waves in weakening the walls and creating high humidity conditions (for example, at the Fort of San José), and the increase in vegetation and rising sea level causing constant flooding (for example in Lazareto and Fort of San José). These observations do not only affect the areas studied but have also had a significant impact on other areas of the city such as El Laguito, Bocagrande Castillo Grande, the historic centre, Getsemaní, San Diego, Marbella, and the surroundings of the Ciénaga de la Virgen (Del Cairo *et al.* 2023a; 2023b).

Figure 5.6: Measurement of affected historical remains by community co-researchers in 2023. Copyright: Gabriela Caro.

As part of this study, we started developing photogrammetric models of certain sites and heritage assets, which will serve as a baseline to initiate a monitoring process using digital technologies to identify the climate change related deterioration observed in maritime cultural heritage of Cartagena through a comparison of the point clouds created. The virtual models were also used to model the future effects of sea level rise on two properties of cultural interest shown in Figure 5.7: the fort of San José in Bocachica and the tilery of San Bernabé in Tierrabomba.

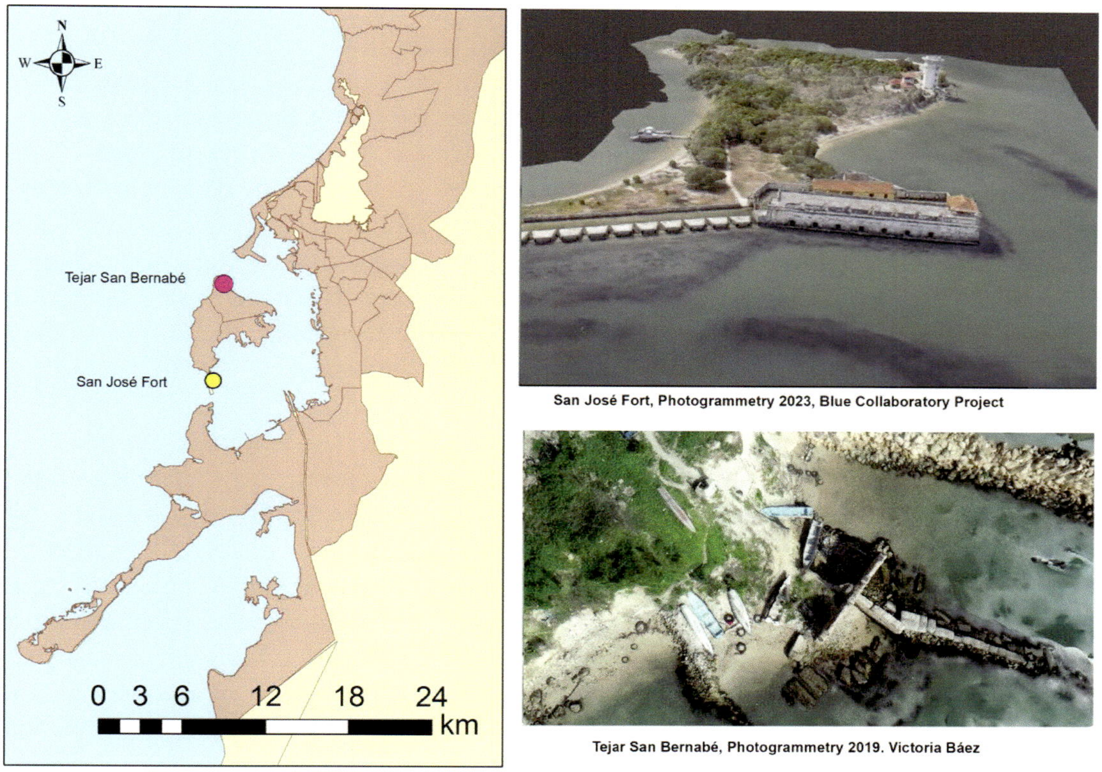

Figure 5.7: Photogrammetric models of San José fort in Bocachica and the tilery of San Bernabé in Tierrabomba, used for the monitoring of archaeological sites in Cartagena de Indias. Copyright: Colaborartorio Azul.

Through this transdisciplinary approach and integration of various data (including narratives and testimonies of local people, climate data models, images and photographs, observations on landscape changes, and maps), it has been possible to generate a first comprehensive understanding and interpretation of the vulnerability Cartagena de Indias' heritage to climate change. The participation of many different actors who are related to cultural heritage and are affected by climate change has been central for that. Thus, the project proposes a working methodology for site monitoring and mitigation measures that integrates local knowledge, together with knowledge of the natural, historical and archaeological research aspects of the project. This could encourage greater ownership of the protection measures and more effective management of vulnerabilities identified, while generating joint action and demonstrating that there can be direct communication and cooperation among the various actors and stakeholders of one city.

Integrating knowledge: reflecting on transdisciplinary approaches to understanding climate change impacts on cultural heritage

The transdisciplinary approach of *Colaborartorio Azul* offered new ways of integrating different data and types of knowledge in heritage research. This fits the observation made by Nicolescu (2002: 44), who explains that 'the transdisciplinary concerns that which is at once between the disciplines, across the different disciplines, and beyond all disciplines. Its goal is the understanding of the present world of which one of the imperatives is the unity of knowledge'. In our experience, the participatory aspects of the project and the involvement of non-academic actors in the research process, noted also by Mauser *et al.* (2013), have been key in achieving unity of knowledge to address a present global but also local challenge. Furthermore, involving non-academic actors in the research provided opportunities for local stakeholders or members of the community to act as agents of local transformation, contributing to the protection of those affected and considering the issues researched for various social and economic activities of the region, such as tourism, that are usually linked to heritage and its value. This is an important conclusion of the project so far, as the lack of knowledge and disarticulation of the significance of the community perspectives, especially on issues such as climate change, frequently results in 'the likelihood of producing maladaptive plans and actions, which may result in irreversible changes or major losses' (Herrera *et al.* 2012: 27). Hence, through the project, it has been possible to build networks of exchange between the public, various stakeholders, and academia with the aim to continuously generate mechanisms for community participation and discussion on significant decision-making related to the protection of heritage from climate change. This follows closely, Meadow *et al.*'s (2015) observation that the co-production of knowledge can be an effective way to approach climate science, as it enables dialogue and collaboration between scientists and decision makers. Similarly, Aktürk and Dastgerdi (2021) state that the most significant challenges for the resilience of cultural landscapes to climate change are institutional, technical, financial, and socio-cultural. For this reason they consider that 'cultural heritage institutions and territorial planners need to develop place-based governance and a participatory adaptation framework at the local level. This type of governance is supposed to consider the local community's needs, their participation in decision making, and its ability to express self-recognition of its identity and heritage' (Aktürk and Dastgerdi 2021: 13).

Another important conclusion of the project so far has been the significance of time, as a factor in implementing transdisciplinarity and integration of knowledge. The project was purposefully designed with a diachronic perspective through three temporal lenses. By examining the challenge of climate change and impacts on heritage through the long-term lens it has been possible to evaluate the long-term environmental changes in Cartagena de Indias from the perspectives of history, geography, archaeology, oceanography and hydrography. The medium-term perspective, on the other hand, has encouraged interaction between more recent historical research, ethnography, anthropology and marine sciences. Particularly interesting have been workshops on topics such photo ethnography, where participants recalled and reflected on specific historical events and they could connect personal and collective stories, creating a rich and complex tapestry of shared experiences. Similarly, during the 'Mandala' workshops (Figure 5.3), it was possible to establish links and support networks between local, institutional, and academic actors to create the appropriate settings for listening and discussion with each other. Finally, through the short-term perspective, the dialogue between researchers from marine sciences and anthropology, and data from citizen science has been initiated, while the creation of dissemination strategies and social networks were prioritised. From this it was possible to carry out pilot measurements of impacts on some heritage sites and establish the variables and categories of analysis to be used in the present and future. This multitemporal approach with the three perspectives, which looked at the past, present, and the future, generated a process of triangular co-participation between institutions/stakeholders, community, and academia that has been very fruitful. The analysis of collected data so far has ensured that temporalities are not confined to the limits of a single discipline. On the contrary, different research techniques can be used and sources of information can be cross-referenced.

However, by its nature, transdisciplinarity also generates a number of complex and multifaceted challenges that must be addressed to achieve effective collaborative work. Mauser *et al.* (2013) mentions several of these including the need to develop new interpersonal skills that are essential to facilitate effective communication and smooth collaboration between different project participants, the requirement to remove persistent inequalities that may exist within the team or in the wider context of the work, ensuring that all members have equal opportunities to contribute to and benefit from the project, the importance of managing change effectively, maintaining an open mind to adapt to new ideas and approaches that may emerge in the course of collaborative work, and clarifying the roles and responsibilities of individual members and partners so as to avoid confusion and duplication of effort, and finally, the development of new support systems that enable transdisciplinary teams to function effectively. In our experience, transdisciplinary work coupled with work with local communities has promoted the inclusion of non-academic members of the public in heritage research. This has generated opportunities for education and training for local members of the public which contributes to reducing inequality gaps in research, especially because maritime and underwater archaeology have been frequently characterised by additional barriers for underrepresented communities that wish to access it, particularly in terms of knowledge, resources and the specific skills required to be included in this type of work (Lemke *et al.* 2022: 2).

Similarly, Meadow *et al.* (2015: 184) also stress the importance of sufficient time on the part of researchers to build the transdisciplinary and co-production networks in order to carry out the various activities required for a project. This is because this type of research is inherently time-consuming and requires sustained commitment over time to achieve meaningful results. In addition, it is important to highlight the challenges imposed by disciplinary language at times. Differences in terminology and conceptual approaches between disciplines and difficulties of the public to follow academic research can create communication barriers that hinder collaboration. To mitigate these problems, in *Colaborartorio Azul* it has been necessary to build specific glossaries to assist in the creation of common categories of analysis, especially for climate change measurements. As explained further in Chapter 6 of this volume, these glossaries facilitated a shared understanding and allowed for a more cohesive and coherent analysis of the data, thus contributing to the success of the transdisciplinary project.

Other studies, such as those by Graham Allsop *et al.* (2017), and Lemke *et al.* (2022), also propose projects that involve communities in transdisciplinarity, referring to their importance in the response to the challenges posed by new contexts, including those encountered with climate change. Our experience with communities in Cartagena de Indias supports this view. The local and indigenous knowledge held by members of the public, as well as the opportunities to monitor spaces of the built environment in the maritime spectrum by communities during the climate emergency opens opportunities for gathering new data and observations while offering initiatives for community-led safeguarding of cultural heritage assets.

Conclusion

Colaborartorio Azul aims at recording how climate change impacts maritime, coastal, and underwater heritage resources for the communities and inhabitants of Cartagena de Indias in Colombia. Following a comprehensive understanding of the maritime cultural landscape of the region, it has been possible to instigate a collaboration among different stakeholders and communities to identify the effects of climate change on tangible and intangible dimensions of the city's heritage and propose adaptation and mitigation measures. Thanks to the different perspectives provided by the training and different disciplinary competencies of the co-researchers, as well as the expertise of the stakeholders and communities involved, along with the lived experiences of the local inhabitants, it has been possible to integrate and create a vibrant dialogue about the past, the present and the future of the city and its heritage. Transdisciplinary and co-production approaches have allowed the recognition of significant transformations of the natural environment in Cartagena de Indias, but also a start in documenting, evaluating and monitoring the effects of climate change on maritime and cultural heritage. In addition, through the different working groups it has been possible to create exchange networks between public, private, and academic actors in the city. Similarly, this project has empowered local communities by fully integrating co-production into the investigative process and giving agency to those most affected by climate change and heritage loss. Thus, by highlighting the importance of community resilience and dialogue, we envision creating effective and sustainable mechanisms for the protection of cultural heritage in the present and future.

Colaborartorio Azul and its methodology have created a low-cost transdisciplinary and co-created understanding of climate change impacts on maritime cultural heritage in Cartagena de Indias and initiated contact and dialogue between significant members of the city. These first results of the project signify a way forward for safeguarding maritime cultural heritage at Cartagena de Indias during the climate emergency, through initiatives that support the close collaboration of academics, communities, and other stakeholders of the city. This might be a solution applicable in the wider region of Latin America, where the need for long-term sustainable projects with limited economic resources has been identified. For this reason, this research has been recognised locally and globally as 'good practice', following the principles of the ICOMOS (2022) project 'Preserving Legacies: A Future for Our Past'. Hence, now Cartagena de Indias has become one of the 10 observation sites worldwide to assess the effects of climate change on its cultural heritage, within the framework organised by National Geographic, ICOMOS, Climate Heritage Network and the University of San Diego, California.

Finally, it should be noted that the results of this first phase of the project are leading to the second phase, which aims to evaluate and monitor the effects of climate change through the use of new technologies, the creation of analysis matrices and a virtual application. In this way, the conceptual and methodological bases for evaluating and monitoring cultural heritage in the coastal or underwater zone of Cartagena de Indias will be put into practice.

References

Alcaldía de Cartagena de Indias 2021. Formulación y adopción del plan integral de gestión del cambio climático territorial - PIGCC-4C- del Distrito de Cartagena de Indias, en el marco de lo dispuesto por la Ley 1931 del 2018. Producto de la Fase I de alistamiento. Plan integral de gestión del Cambio climático (Plan 4C) Cartagena compatible con el clima. Report.

Aktürk, G. and A. S. Dastgerdi 2021. Cultural landscapes under the threat of climate change: A systematic study of barriers to resilience. Sustainability 13.17: 9974. <https://doi.org/10.3390/su13179974>

Afanador, F., J. Gómez, F. Orozco and A. Carvajal 2008. Ortofotografía digital y datos LiDAR para el control y administración del litoral de la isla de Tierrabomba, Caribe Colombiano. Cartagena de Indias, Colombia. *Boletín Científico CIOH* 26.

Barba Meinecke, H., B. Díaz Fuentes and P. Luna Erreguerena 2010. El patrimonio arqueológico en costas y mares de México ante el Cambio Climático, in E. Rivera-Arriaga, I. Azuz- Adeath, L. Alpuche Gual and G.J. Villalobos-Zapata (eds) *Cambio Climático en México un Enfoque Costero-Marino*: 639-654. Campeche: Universidad Autónoma de Campeche, Cetys-Universidad, Gobierno del Estado de Campeche.

Brito Martínez, O. B., E. A. S. Bueno and G.R. Artiles 2023. *Indicadores para medir impactos del cambio climático en el patrimonio histórico cultural*. Cuba: Medio Ambiente y Desarrollo.

Climate Science Alliance – Descendants United for Nature, Adaptation, and Sustainability (DUNAS), viewed 10 October 2025, <https://www.climatesciencealliance.org/dunas>.

Colaborartorio Azul, viewed 25 October 2025, <https://colaborartorioazul.weebly.com/>.

Cook, I. 2019. Climate Change and Cultural Heritage: developing a landscape-scale vulnerability framework to measure and manage the impact of climate change on coastal historic landscapes. Unpublished PhD dissertation, University of Sheffield.

Del Cairo Hurtado, C., G. Caro, L. Rozo, L. V. Báez, D. Acosta, N. Palacios, J. Sarmiento and J.A. Aldana Mendoza 2023a. Proyecto Colaborartorio Azul. Efectos Del Cambio Climático En Los Bienes De Interés Cultural De Cartagena De Indias, Colombia. Bogotá: Fundación Anfibia. Report.

Del Cairo Hurtado, C., G. Caro León, G.L. Hernández Zárate, L. P. Rozo Pinzón, S. E. Vallejo Quintero, J. A. Aldana Mendoza, J. Cuta Jiménez and L. V. Báez Santos 2023b. Identificación coparticipativa de los impactos derivados del cambio climático sobre los patrimonios culturales marítimos en la isla de Tierra Bomba, Cartagena de Indias, Colombia. *Boletín Científico CIOH* 42.2: 7–25.

Del Cairo, C., G. Caro, C. Riera, L. Rozo and J. A. Aldana Mendoza 2024. Colaborartorio Azul Fase II: Propuesta de Evaluación y Monitoreo Co-Participativo sobre los Efectos del Cambio Climático en el Patrimonio Cultural Costero de Cartagena de Indias, Colombia. Universidad de Cartagena - Instituto de Patrimonio y Cultura de Cartagena de Indias (IPCC) - Fundación Colombia Anfibia. Report

Ezcurra, P. and I. Rivera-Collazo 2018. Una evaluación de los impactos del Cambio Climático en el Patrimonio Cultural de Puerto Rico con un estudio de caso sobre el aumento del nivel del mar. *Revista de Patrimonio Cultural* 32: 198–209.

Graham Allsop, E. L., J. Hambly and T. C. Dawson 2017. Scotland's eroding heritage: a collaborative response to the impact of climate change. *Archaeological Review from Cambridge* 32.2: 141-158.

Herrera, M. C. P., A. Rueda AND C. Pinzón 2012. Percepciones sobre los fenómenos de variabilidad climática y cambio climático entre campesinos del centro de Santander, Colombia. *Ambiente y Desarrollo* 16.31: 25–37.

ICOMOS, 2020, ICOMOS Declares a Climate Emergency, viewed 25 October 2023, <https://www.icomos.org/en/focus/climate-change/85740-icomos-declares-a-climate-emergency>.

ICOMOS, 2022, Preserving Legacies: A Future for Our Past, viewed 25 October 2023, <https://www.icomos.org/projet/preserving-legacies-a-future-for-our-past/>.

Kimmell, K. and A. Markham 2021. Opening Keynote Address – The Importance of Place and Heritage in Achieving Climate Solutions", in R. Rushfield (ed.) *Stemming the Tide: Global Strategies for Sustaining Cultural Heritage through Climate Change*: 1–12. Washington, D.C: Smithsonian Institution Scholarly Press. <https://doi.org/10.5479/si.14750727>

Mauser, W., G. Klepper, M. Rice, B. S. Schmalzbauer, H. Hackmann, R. Leemans and H. Moore 2013. Transdisciplinary global change research: the co-creation of knowledge for sustainability. *Current Opinion in Environmental Sustainability* 5.3–4: 420-431. <https://doi.org/10.1016/j.cosust.2013.07.001>

Meadow, A. M., D. B. Ferguson, Z. Guido, A. Horangic, G. Owen, and T. Wall 2015. Moving toward the deliberate coproduction of climate science knowledge. *Weather, Climate, and Society* 7.2: 179–191. <https://doi.org/10.1175/WCAS-D-14-00050.1>

Nicolescu, B. 2002. *Manifiesto of Transdisciplinarity*. Albany: State University of New York Press.

Lemke, A., N. B. Grinnan and J.V. Haigler 2022. Getting Your Feet Wet: Barriers to Inclusivity in Underwater Archaeology and How to Break Them. *Advances in Archaeological Practice* 10.2: 129–139. <https://doi.org/10.1017/aap.2021.34>

Reeder-Myers, L.A. 2015. Cultural Heritage at Risk in the Twenty-First Century: A Vulnerability Assessment of Coastal Archaeological Sites in the United States. *Journal of Island and Coastal Archaeology* 10.3: 436–445. <https://doi.org/10.1080/15564894.2015.1008074>

Ricaurte Villota, C., M.E. Gonzélez Arteaga, O. Coca Dominguez, H.M. Bejarano Espinosa, D.F. Morales Giraldo, C.X. Correa Rojas and M.E. Arteaga 2018 Amenaza y vulnerabilidad por erosión costera en Colombia: enfoque regional para la gestión del riesgo, Santa Marta, INVEMAR.

Rivera-Collazo, I. 2022 Environment, climate and people: Exploring human responses to climate change. Journal of Anthropological Archaeology 68: 101460. <https://doi.org/10.1016/j.jaa.2022.101460>

Rubio-Ardanáz, J. A. 2014. Maritimidad, un concepto y un marco para la comprensión. *Antropología y maritimidad. Entramados y constructos patrimoniales en el Abra y Ría de Bilbao*: 27-43. Spain, Museo Marítimo Ría de Bilbao.

Sánchez, J. A. A., and J D.V. Cardona 2017. Interdisciplinariedad y transdiciplinariedad en la ciencias sociales y humanas: convergencias entre el giro lingüístico y la obra de Pierre Bourdieu. *Kénosis* 5.9: 40–60.

UNESCO, Port, forteresses et ensemble monumental de Carthagène, viewed 20 May 2024, <https://whc.unesco.org/fr/list/285/>.

UNESCO, 2014, Action Plan for the Caribbean: Action Plan for World Heritage in the Caribbean 2015 – 2019, viewed 10 October 2025, <https://whc.unesco.org/es/actividades/846/#PAC>.

UNESCO, 2022, Climate Change and World Heritage, viewed 20 May 2024, <https://whc.unesco.org/en/climatechange/>.

UNESCO The Criteria for Selection, viewed 10 October 2025, <https://whc.unesco.org/en/criteria/>.

UN Sustainable Development Goals, United Nations, Department of Economic and Social Affairs, Sustainable Development, viewed 15 September 2024, <https://sdgs.un.org/goals>

Westerdahl, C. 1992. The maritime cultural landscape. *International Journal of Nautical Archaeology* 21.1: 5–14. <https://doi.org/10.1111/j.1095-9270.1992.tb00336.x>

Westerdahl, C. 2011 The Maritime Cultural Landscape, in B. Ford, D.L. Hamilton and A. Catsambis (eds) *The Oxford Handbook of Maritime Archaeology*: 735–762. Oxford: Oxford University Press.

6.
Co-production and Participatory Approaches in Understanding the Impacts of Climate Change on Maritime Cultural Heritage in Tierrabomba, Cartagena de Indias (Colombia)

Gabriela Caro León, Liliana Rozo Pinzón, Juan Sarmiento Rodríguez, Fernando Cadena Duque

As part of the *Colaborartorio Azul* project of the NGO *Fundación Colombia Anfibia* (see Chapter 5), a local network was proposed to support further cooperation, co-production and active participation of community members and stakeholders, which could serve as the central axis for generating knowledge about climate change and its impacts on cultural heritage on the island of Tierrabomba (*Isla de Tierrabomba*), in Cartagena de Indias, Colombia. This network aimed to offer a suitable framework for interested parties to understand the importance of the local and regional heritage of Cartagena de Indias and work out possible ways of protecting it. The initiative sought to build comprehensive knowledge about the threats and challenges faced. Drawing on diverse perspectives, knowledge is built through engagement events with local populations to identify and recognise the effects of climate change on cultural heritage assets. This approach enhances the community's decision-making capacity and increases opportunities for knowledge exchange through public mechanisms that address the current global climate emergency. Hence, the *Colaborartorio Azul* project has been implementing community, governmental, and academic collaborations and is looking at the risks posed to the city's natural and cultural heritage.

Introduction

In the Caribbean region, the effects of climate change are evident in both marine and terrestrial environments. In marine settings, these effects include ocean water warming, sea-level rise (ranging from 0.2m to 0.9m), and an increase in events such as hurricanes and tropical storms. Additionally, marine ecosystems are impacted by coastal erosion, ocean acidification, and coral bleaching. In terrestrial environments, projected climate change impacts include changes in ecosystems and natural predation cycles, increased precipitation (associated with heightened runoff and its implications), and species migration (Scantlebury 2018).

Several studies, including those by Andrade (2008), Rangel and Montealegre (2003), Pabón (2003), Pabón and Lozano (2005), and Torres and Tsimplis (2013), have analysed the trends in sea level rise along Colombia's coasts. These studies, based on historical data of average sea levels and oceanic topography, conclude that between 1961 and 1990, the sea level rose by 3mm to 5mm annually in the Pacific Ocean and by 1mm to 2mm annually in the Caribbean at the Atlantic Ocean. In Cartagena de Indias, Colombia, however, the rise has been even more alarming, with a rate of 5.3mm per year over the past 44 years—two to three times faster than in nearby ports (Torres and Tsimplis 2013).

The cultural heritage of Cartagena de Indias, located in a maritime coastal context, is predominantly influenced by its natural environment, which determines its state of conservation over time. Climatic conditions, which continuously affect both the tangible and intangible heritage of communities, influence the assigned heritage significance and meanings constructed within the region. Climate change has negatively impacted the development, conservation, and protection of the heritage of maritime and coastal communities (IPCC 2022). For this reason, gaining full understanding of the concepts of climate change, heritage, and community, as well as their interrelations, is considered essential to safeguard, protect and reduce levels of deterioration of cultural heritage during the climate emergency (Rivera-Collazo and Declet Pérez 2017).

After briefly presenting the methodology used and the main objective of this study, the paper lays out the approaches implemented with the communities and stakeholders on the island of Tierrabomba and explains the community-based monitoring carried out through periodic records taken at selected heritage sites and everyday spaces, such as homes. The overall aim has been to build capacity and increase preparedness for the populations of Cartagena de Indias in the face of the climate emergency.

Methodology and objectives

To understand the impacts of climatic phenomena from an interinstitutional and community perspective, the *Colaborartorio Azul* project implemented the following methodology that combined quantitative and qualitative data, structured into three main components with different temporal scales of analysis.

First, to investigate climatic and landscape changes of the last 300 years, primary historical sources, such as cartographic data and travellers' maps, were utilised to identify and analyse natural elements, characterise the climate, and document changes to the built environment and heritage over time. Second, practical activities were conducted with local communities using tools such as field journals, audiovisual recordings, social cartography, and participatory activities. All these facilitated the reconstruction of a 'landscape memory' spanning the last two centuries, integrating perceptions of climate change with the physical environment and traditional community practices, including local knowledge from the 1960s to the present. Finally, practical monitoring activities were developed to cover the contemporary period, documenting climatic conditions and establishing specific points in the city to measure and monitor territorial changes and transformations associated with climatic variations (Del Cairo *et al.* 2022).

The main objective of this paper is to present one of the outcomes of the work, which is still ongoing. This is the process used for the identification of key stakeholders, and the research that allowed the evaluation of risks associated with threats to heritage arising from changes in environmental variables affecting the study area (Del Cairo *et al.* 2023).

Identifying key actors, resources, potentials and risks

Considering the participatory focus of *Colaborartorio Azul*, an outreach initiative was organised for the local communities in three areas of the island: Bocachica, Tierrabomba, and Caño del Oro (Figure 6.1). The objective was to identify and map all key stakeholders related to maritime cultural heritage and climate change, categorise them into interest groups and, therefore, promote maximum participation in the implementation of the pilot project (Del Cairo *et al.* 2022).

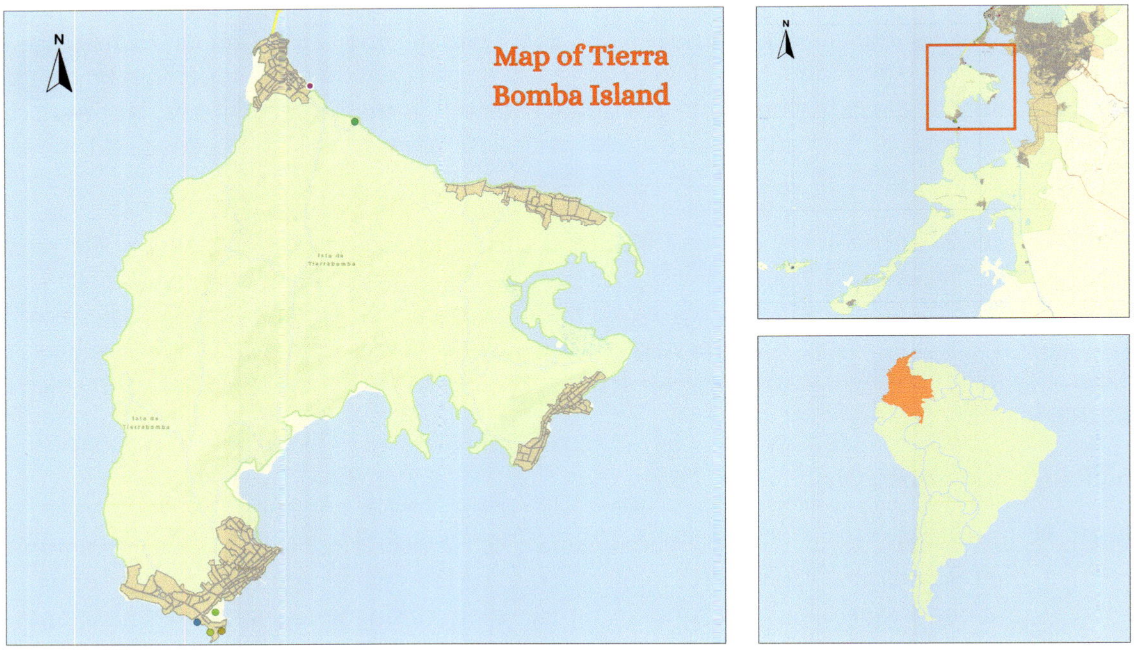

Figure 6.1: General location of the Colaborartorio Azul project, Tierrabomba Island, Cartagena de Indias. Copyright: Colaborartorio Azul.

To identify the key stakeholders, the territorial context of the community was considered, working from three perspectives: sociocultural, institutional, and scientific. These approaches align with the guidelines of the International Centre for the Study of the Preservation and Restoration of Cultural Property (ICCROM 2015) on people-centred approaches to cultural heritage conservation.

A study of the island's cultural heritage assets was conducted to identify potential risks and environmental threats. This characterisation enabled the mapping of resources and sites and the prioritisation of those at greatest risk. Based on this, the objectives of the initiative were grouped according to the need to contribute to the sustainability of cultural heritage, and recognised communities and heritage networks as significant actors with a strong sense of participation in cultural identity surrounding heritage sites. Professionals or academics were considered for their access to knowledge through research and primary data collection and analysis, as well as their capacity to build a support base for heritage conservation. Similarly public institutions responsible for decision making and political ideas were considered as key to creating policy opportunities and articulating heritage-centred cultural planning and public engagement (Del Cairo *et al.* 2022; 2023). Finally, communities and residents of the region were considered for their significant local knowledge, along with their empirical understandings of landscape, heritage, and long-term changes happening in the areas where they live and work.

The result was 28 stakeholders that were identified and invited to participate in the project (Table 6.1). The categorisation of these gave us an opportunity to define each group's influence in decision making—whether as public or private organisations—and their impact at local, regional, or international level. Based on all the above factors, their potential for participation in actions and discussions related to climate change and cultural heritage was determined (Del Cairo *et al.* 2022).

Table 6.1. Stakeholders identified in Cartagena de Indias.

Stakeholder	Type of organisation	Ownership (Private, Public, NGO, etc.)	Interest in Impacts of Climate Change on Maritime Cultural Heritage
CULTURE MINISTRY	GOVERNMENT	PUBLIC	HIGH
MINISTRY OF ENVIRONMENT	GOVERNMENT	PUBLIC	MEDIUM
IDEAM	INSTITUTION	PUBLIC	MEDIUM
MAYOR OF CARTAGENA	GOVERNMENT	PUBLIC	MEDIUM
INSTITUTE OF CULTURAL HERITAGE OF CARTAGENA (IPCC)	INSTITUTION	PUBLIC	HIGH
WORKSHOP SCHOOL OF CARTAGENA DE INDIAS	INSTITUTION	PUBLIC	HIGH
UNIVERSITY CARTAGENA	ACADEMIC	PUBLIC	LOW
TECHNOLOGICAL UNIVERSITY OF BOLÍVAR	ACADEMIC	PUBLIC	MEDIUM
CONSERVE GROUP	INSTITUTION	PRIVATE	MEDIUM
CORALES DEL ROSARIO AND SAN BERNARDO NATIONAL NATURAL PARK	ACADEMIC	PUBLIC	LOW
CARTAGENA PUBLIC IMPROVEMENTS SOCIETY	INSTITUTION	PRIVATE	LOW
RISK MANAGEMENT UNIT (IDU)	INSTITUTION	PUBLIC	LOW
COLOMBIAN INSTITUTE OF ANTHROPOLOGY AND HISTORY (ICANH)	INSTITUTION	PUBLIC	HIGH
NATIONAL NAVY OF COLOMBIA (ARC)	INSTITUTION	PUBLIC	HIGH
GENERAL MARITIME DIRECTORATE (DIMAR)	GOVERNMENT	PUBLIC	HIGH
CAPTAINTY OF PORT OF CARTAGENA	INSTITUTION	PUBLIC	HIGH
DEPARTMENT OF DIVING AND RESCUE (DEBUSA)	INSTITUTION	PUBLIC	HIGH
CARIBBEAN OCEANOGRAPHIC AND HYDROGRAPHIC RESEARCH CENTER (CIOH)	INSTITUTION	PUBLIC	HIGH
COLOMBIAN OCEAN COMMISSION (CCO)	INSTITUTION	PUBLIC	HIGH
CARIBBEAN NAVAL MUSEUM	MUSEUM	PUBLIC	MEDIUM
LOCAL COMMUNITY ACTION BOARDS	COMMUNITY	PUBLIC	MEDIUM
CARTAGENA DIVERS	BUSINESS	PRIVATE	MEDIUM
BARU DIVERS	BUSINESS	PRIVATE	MEDIUM
TIERRABOMBA COMMUNITY MUSEUM	MUSEUM	PUBLIC	LOW
KAREX WATCHERS	COMMUNITY	PUBLIC	HIGH
BENKOS BIOHÓ HOUSE OF CULTURE	COMMUNITY	PUBLIC	LOW
BAHIA FOUNDATION	COMMUNITY	PRIVATE	LOW
THE NATURAL PARK OF THE GULF OF MORBIHAN - CACTUS	INSTITUTION	PRIVATE	MEDIUM

The following classification of stakeholders was based on their interest and capacity to support the project:

High: Stakeholders with significant interest in participating in the project's implementation and the ability to contribute financial and human resources.

Medium: Stakeholders interested in the project but without sufficient capacity to provide financial or logistical support.

Low: Stakeholders with neither interest nor resources to contribute.

The categorisation of key stakeholders played an important role in the implementation of the project's co-production and participatory approaches. The primary identification and classification of stakeholders into interest groups enabled the highest possible level of participation in our pilot study. Furthermore, subgroups were also identified within these results. For example, categorisation into private or public initiatives and recognition of the type of organisation (namely, government, independent institute/body, business or community) helped us map the available expertise, human resources, funding and availability for capacity building in the region. This became even clearer in a further classification when the interests of stakeholders in relation to the project's objectives were defined, and their potential to support the project's development was evaluated based on their capacities. Subsequently, a call was issued for participation in co-production workshops, which provided valuable input for assessing the effects of climate change in the maritime-coastal context represented in *Isla de Tierrabomba* (Del Cairo *et al.* 2022, 2023).

Assessment of the effects of climate change in coastal heritage contexts

The second phase of our study involved implementing a participatory process to strengthen the theoretical and practical capacities of the community, aimed at identifying, evaluating, and measuring the impacts of climate change on the built environment and heritage assets of *Isla de Tierrabomba*. To achieve this, a theoretical-practical workshop was designed, allowing participants to acquire specific knowledge from experts and apply it in practice, thereby consolidating the information gained.

Initially, a systematisation of terms and/or concepts was conducted with the stakeholder partners participating to establish clear indicators for assessing the heritage risks associated with climate change. The community worked with concepts such as 'vulnerability', 'threat', and 'impact', which facilitated the definition of the 'fragility' or 'resilience' of cultural assets in the face of climate-related threats (Del Cairo *et al.* 2023). Among the concepts discussed was the definition of climate change as established by Law 1523 of 2012 in Colombia: 'A significant statistical variation in the average state of the climate or its variability, which persists for a prolonged period (normally decades or longer)'. It was also determined that climate change could be 'directly or indirectly attributed to human activities that alter the global atmospheric composition, and which adds to the natural climate variability observed over comparable time periods', as also explained by UNESCO (2014). Another relevant concept was that of 'risk', defined as 'the damage or potential losses that may occur due to hazardous physical events of natural, socio-natural, technological, bio-sanitary, or unintentional anthropogenic origin, over a specific period, determined by the vulnerability of the exposed elements'. Disaster risk arises from the combination of threat and vulnerability, as outlined in Law 1523 of 2012. From this perspective, risk is understood as an extreme natural phenomenon that exceeds normal conditions, has the potential to affect a system, and exposes its elements to the capacity to assimilate, endure, and recover from the phenomenon.

Using the proposed terminology, a collaborative exercise was carried out in our workshops to identify and evaluate the known effects of climate change on cultural heritage assets of the *Isla de Tierrabomba* as well as the capacity of various actors and stakeholders in the city to deal with them. This process consolidated a matrix consisting of a series of actions that could link threats with potential solutions in a hyperlocal scale that could work for Cartagena de Indias. This matrix and associated series of actions are outlined below:

a) Identifying Vulnerabilities:

- Institutional Value: Evaluate the capacity of local and national institutions to manage and protect cultural assets.
- Local Value: Determine the cultural and social significance of heritage sites for local communities.
- Commercial Value: Analyse the economic impact of heritage sites loss on tourism and other economic and commercial activities.
- Regulatory Framework: Review existing policies and regulations that protect cultural assets.
- Site Usage: Examine how heritage sites are used and the frequency of their usage.
- Environment: Consider the natural and built environments surrounding the cultural heritage assets.

b) Threat Assessment:

- Natural Threats: Identify natural threats, such as sea-level rise, coastal erosion, storms, and ocean acidification.
- Cultural Threats: Identify human factors such as urban development, pollution, and uncontrolled tourism that could out heritage assets at risk.

c) Characterising Alterations:

- Cracks and Deformations: Identify and document cracks and deformations in heritage structures.
- Material Losses and Landslides: Record material losses and landslides that compromise the integrity of cultural heritage assets.
- Colour and Surface Alterations: Evaluate changes in material colour and surface due to exposure to climatic factors.
- Bio-colonisation: Analyse the existence of biological organisms (local or invasive) that may deteriorate heritage materials.

d) Data Classification and Analysis:

- Data Collection: Use technological tools and participatory methods to collect data on observed impacts.
- Risk Analysis: Apply risk assessment methodologies to identify the most vulnerable assets.
- Impact Mapping: Create detailed maps to visualise the areas and cultural assets most affected by climate change.

e) Developing Mitigation and Adaptation Strategies:

- Participatory Planning: Engage local communities in planning decision making, develop and execute mitigation and adaptation strategies.
- Implementation of Measures: Apply specific actions to reduce vulnerabilities and protect cultural heritage assets.
- Monitoring and Evaluation: Establish a continuous monitoring and evaluation system and adapt strategies based on the results.

Figure 6.2: Development of the project website. Copyright: Colaborartorio Azul.

From the above process and the practical workshop activities associated with it, it became obvious that evaluating the effects of climate change on coastal contexts requires collaboration between multiple actors, including local communities, academic institutions, government entities, and international organisations. Only through a participatory and multidisciplinary approach can effective solutions be developed and applied to preserve cultural heritage in the face of the current climatic challenges (Del Cairo *et al.* 2023).

In the final phase of the study, a dissemination strategy was developed and implemented to communicate the project's results and suggestions, aiming to define a target audience and to increase the project's visibility online. The goal was to increase visibility of the project's website (Figure 6.2) and social media platforms, ensuring a sustained legacy of the project activity and new interactions with digital users. Local populations, actively co-creating content and promoting community-driven activities on social media, were the primary audience (Figure 6.3). At a secondary level, government entities were engaged due to the pilot project's impact on communities—a model that can be replicated in other regions of the country, but also abroad.

To promote the safeguarding of cultural heritage in Tierrabomba, Bocachica, and Caño del Oro on Isla de Tierrabomba, practical sessions were initiated to foster continuous knowledge exchange, enabling the implementation of transdisciplinary, multisectoral, and interinstitutional actions to address this phenomenon. As will be explained below, the local community, as part of the co-production team, played a key role in demonstrating how rising tides and strong storms with dynamic winds and currents in recent years have gradually caused the loss of historical and archaeological sites in Cartagena de Indias. This helped to prove how coastal erosion, structural cracking of built heritage, the formation of biological crusts, and flooding by seawater have been damaging heritage assets.

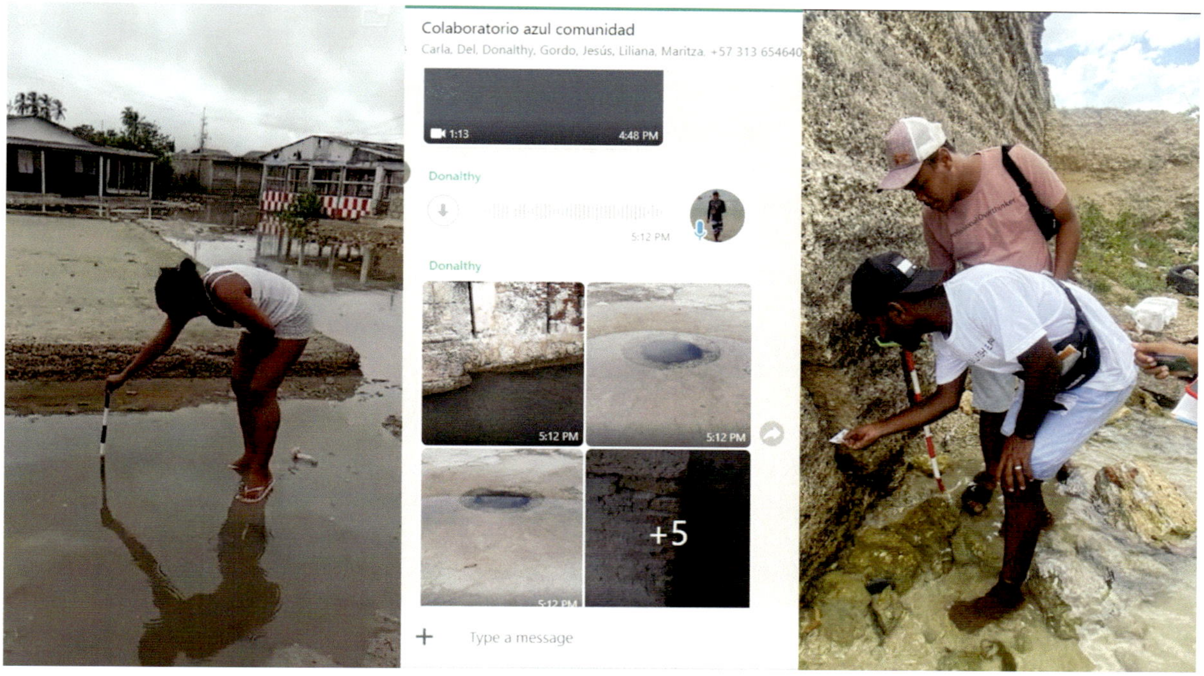

Figure 6.3: Practical sessions carried out by members of the community to monitor impacts due to climate change in the community of Bocachica. These were shared on social media by the project but also the individuals participating. Copyright: Del Cairo et al. 2023.

Measuring the impacts of climate change in practice

The participation of local populations has been fundamental in implementing monitoring matrices and identifying the impacts of climate change in Cartagena de Indias. In this context, the resilience of cultural heritage is conceived through the support networks it generates within the local community, driven by the solidarity of those who share a set of cultural goods and practices, as pointed out previously by García Canclini (1999). Therefore, community participation and solidarity in decision making is essential to energise cultural heritage processes, addressing the challenges and fostering resilience aimed at identifying, evaluating, and measuring the impacts of climate change on the heritage of *Isla de Tierrabomba*.

Based on this premise, a theoretical and practical workshop was conducted with members of the local community of Cartagena de Indias and *Isla de Tierrabomba*. During this, key stakeholders in local districts, previously identified as described above, were contacted to take part in the pilot project through a socialisation session. This aimed at encouraging their active and strategic participation in the project's development and practical actions, and also adopting the results and recommendations of the project after its completion. The workshop started with an introductory session led by professionals specialising in climate change and heritage, providing general insights into the impact of climate change on maritime cultural heritage. Subsequently, the workshop facilitated discussions on the key elements and climatic factors currently affecting Cartagena de Indias and *Isla de Tierrabomba* (Del Cairo *et al.* 2022). Representatives from various organisations and communities participated in the events. Some of these were:

- University of Cartagena: Carlos Acevedo
- Escuela Taller de Cartagena de Indias (ETCAR): Rodrigo Alfaro
- Fundación Vigías de Carex: José Gregorio Julio, Justo Germán Castro, Reynaldo Julio Babilonia
- Instituto de Patrimonio Cultural de Cartagena (IPCC): Luis García, Mariana Carulla Arreaza
- Community members from Tierra Bomba: Marais Rodríguez, Esteisi Córdoba Morales
- Community members from Bocachica: Donayro Guzmán, Álvaro Barrera
- Dirección General Marítima (DIMAR): Captain Gina Hernández
- Naval Museum of the Caribbean: Gonzalo Zúñiga, Marlín Romero
- Fundación Los Jagüeyes: Martín Raúl González

The main activity of the workshop began with a brief presentation on the *Colaborartorio Azul* project. The project's objectives were outlined, along with some theoretical frameworks related to climate change and its effects on maritime and coastal cultural heritage. Additionally, historical perspectives on coastal changes were introduced, featuring specific examples from Cartagena de Indias to foster discussion about measurable impacts in the local environment. For example, a historical analysis of coastal changes in the region demonstrated how Cartagena's coastline has progressively receded over time, indicating that the city is sinking. In particular, the area of Bocachica has lost up to 254m of coastline. Following that, a practical group activity was carried out using newspaper clippings from the past fifty years that documented climate-related impacts observed in Cartagena de Indias. This activity employed the methodology of photo ethnography, which evokes collective memory through images and narratives (Berger and Mohr 2009). During this, participants worked with photographs and newspaper headlines illustrating the city's environmental challenges over the decades, including phenomena such as strong winds, rains, floods, and coastal erosion. The primary goal was to explore how participants have experienced and perceived these climatic changes over time.

From the discussions generated during the workshop, six key aspects emerged:

- Climate change
- Impacts on the community
- Maritime-coastal environment
- Heritage
- Climatic variables
- Units of observation for the impacts

The aim of the workshop to initiate discussions through recollections or with the use of the historical archival newspaper material was accomplished. Topics such as floods, rainfall, cyclones, storms, and other impacts on coastal areas and heritage sites were addressed. Discussion groups were formed to analyse and interpret the data further. Each group reviewed the selected newspapers and results of the group work and then engaged in a debate on the topics covered.

Overall, the workshop was successful in enabling the identification of climate change processes through historical materials, which highlighted the environmental transformations on the island. Additionally, the hands-on activities enabled the participants to understand experientially and connect emotionally to the theoretical content of local climatic history. Additionally, the workshop implemented methods for integrating cultural memory and the experiences of various population groups, such as older adults, single mothers, and fishermen. This approach sought to understand existing perceptions, biases and landscape experiences of the local residents, and also how the way of inhabiting a region defines knowledge and identities. Oral histories, visual data, and other local knowledge information were also collected during the workshop. Further photography, video interviews, field journals, and more participatory workshops were encouraged and organised after the first pilot event (Del Cairo *et al.* 2022).

The data collected through these workshops and engagement activities identified several variables associated with the perception of climate impact, reflecting the community's experience of interacting with the natural environment. This perspective integrates how 'individuals perceive, appropriate, and interpret the climatic meteorological events occurring around them; the concept of climate is a cultural construct developed from material and symbolic processes that reflect cultural, spatial, and historical dimensions' as described by Forero *et al.* (2014). Within the community, the prevailing memories and perceptions involved intense rains, winds, and floods linked to new precipitation patterns and temperature variations that transform their familiar landscape, built heritage, and heritage sites, accelerating deterioration processes. Concurrently, other threats associated with diseases were observed. Indeed, a correlation exists between climatic variables, environmental deterioration, and health conditions due to waste, garbage dumping, and pollution from residues from daily and industrial activities, which affect water conditions and negatively impact the quality of life for those living on the coast (Del Cairo *et al.* 2023).

Based on these findings, an initial roadmap was established for creating strategies to potentially implement further adaptation and safeguarding measures. Efforts focused on assisting communities to understand climatic changes, weather variables, and natural environment changes, such as temperature fluctuations, changes in precipitation, wind variations, floods, landslides, sea level rise, and ocean properties among others. The creation of collaborative definitions of climate change impact indicators on maritime and underwater tangible cultural heritage from a local perspective turned out to be highly effective in helping locals to understand the risks and potential solutions. Because of this, it was proposed that project participants and workshop leaders could work together in the future to develop measurement instruments, control points, indicators, and other analytical methods that can empower local residents to realise and monitor the threat scale for heritage sites in their area, which was trialled in the project (Figure 6.3). These efforts would consider closely the participants' daily lives and their knowledge of the dynamics and components of the landscape existing on the island (Del Cairo *et al.* 2023). Notably, the most frequently mentioned archaeological sites by the workshop participants included the San Felipe, San José (Figure 6.4), and San Fernando fortifications, along with the San Bernabé kiln. On a secondary level, the Santa Bárbara and Santiago batteries were identified as significant landmarks, while submerged heritage and coastal museums were mentioned in more general terms. From these discussions, it also became clear that urban areas, such as the Cartagena city centre and Getsemaní, were considered among the most affected locations, along with rural areas in Bocachica, Tierrabomba, and Barú. Members of the community could recollect that climatic changes and environmental phenomena, such as sea level rise and coastal erosion, have already caused significant loss of archaeological contexts and severe deterioration of heritage assets in those areas.

Figure 6.4: Photographic archives of the Fuerte de San José archaeological site between 2017 and 2022. Copyright: Colaborartorio Azul.

Specific methodologies were implemented to continue monitoring the impacts of climate change on cultural heritage assets beyond the time of the pilot study. This task was one of the most significant aspects of the project, as the effects can vary depending on the natural context and the specific social and environmental conditions of each site. Therefore, the processes of recording, measuring, and constant monitoring, carried out by community members, proved to be fundamental and highly effective (Figure 6.3). These activities not only generated valuable data for other aspects of *Colaborartorio Azul,* but also added to the theoretical and practical skills gained by the participants during the workshops. To facilitate measurement and monitoring by the community and the identified key stakeholders with capacity in doing so, recording instruments were designed in collaboration with the project participants. These included markers made from simple materials, such as PVC pipes, measuring tapes, and scales. Additionally, audiovisual tools such as cameras (free-standing or on mobile phones) and equipment for audio recordings were incorporated, along with analogue methods of documentation such as drawings, diagrams, and field journals.

To foster discussion and knowledge exchange about the records created by the participating stakeholders and communities, a WhatsApp group called 'Climate Change Cartagena' was created (Figure 6.3). This digital space brought together project researchers, enabling the wider communication and integration of diverse actors, and, thus, creating a strong support network.

These actions were essential in advancing to the final stage of the project, which aimed at promoting effective collaboration and consolidating the lessons learned and the results achieved. For example, in the project's final stage, a communication strategy was developed to disseminate our proposed guidelines for monitoring climate change impacts on maritime and coastal heritage assets of Cartagena de Indias. The communication strategy was divided into two phases. The first phase focused on a dissemination plan, during which results were evaluated and measured to ensure alignment with the project's objectives. Identifying a target audience, within the local community, that resonated with the project's main goals was crucial for the success of this phase. In a second phase, efforts were directed toward government organisations and the general public to amplify the impact and raise awareness of the challenges posed by climate change to cultural heritage.

In more detail, the project's communication and dissemination strategy focused largely on identifying target audiences that could ensure the lasting impact and legacy of our pilot study. Given the importance of the local population, we aimed at actively involving them in content creation and promotion of their activities on social media (Figure 6.3). Through posts and activities on a variety of digital platforms, we managed to increase traffic to the project website, ensuring the local community stayed informed virtually about the project's progress, and also to instigate new interest in our work and achievements. Additionally, efforts were made to engage government organisations, recognising the potential impact of our pilot project on community-driven cultural heritage management in the long term, and its replicability in other areas of the country. Moreover, the strategy sought to raise awareness among the general public about the connection between climate change and shared heritage.

To implement this strategy further, the second phase of the project included the launch of the website 'Participatory Identification of Climate Change Impacts on the Cultural Heritage of Cartagena de Indias, Colombia' that broadened and encouraged participation (*Colaborartorio Azul* n.d.). The site featured a sitemap, links to the project's social media, and the project's logo, all designed to facilitate navigation and foster interaction. A dedicated space was also created to engage and showcase the stakeholders involved, including local communities, regional and local governments, and other interested parties and organisations who wished to contribute to the project. All the above actions ensured the continuity of the community- and stakeholder-driven activities beyond the resources and time of our pilot project, leaving a significant legacy that can help protect maritime cultural heritage and local people from climate change in the future.

Conclusion

Climate change is one of the most critical global challenges of our time. Historical and desk-based analysis conducted during this project clearly demonstrates that Cartagena's coastline has receded significantly over the past 100 years—far more rapidly than in previous centuries. This retreat exposes the city's coastal and low-lying areas to systematic and accelerated flooding at rates faster than the global average. This situation results from rising sea levels caused by global warming and land subsidence, compounded by urban and tourist activities in these areas. These circumstances highlight the urgent need to implement mitigation strategies to address climate change impacts on Cartagena, prevent further land loss, and protect the coastal cultural heritage that is integral to the city's identity. This requires the implementation of collaborative actions that integrate transdisciplinary, multisectoral, and interinstitutional approaches to drive and carry out informed decision making.

Colaborartorio Azul is an exemplary project that has utilised co-production and participatory approaches to build capacity for identifying and monitoring the impacts of climate change on the cultural heritage of Cartagena de Indias, a cornerstone of the city's identity. Following the identification of key stakeholders, including members of the community, this project has been using historical analysis and sustainability education to inform on the key concerns surrounding the effects of climate change on Cartagena's cultural heritage, giving an opportunity to local communities to safeguard and protect their heritage assets for the future.

Community and stakeholder perceptions have revealed critical details and observations on climate factors impacting the city, such as tides, water currents, winds, precipitation, sea-level rise, and high temperatures. These interconnected factors create cascading effects on the environment and the quality of life of local communities. Understanding the interplay between environmental variables, various stakeholders, communities, and cultural heritage has also proved essential for their protection and safeguarding of the built environment, heritage, and the local populations in the long-term. This study underscores the importance of co-production and participatory research in emphasising that local communities possess valuable insights into the effects of climate change and can play a vital role in monitoring impacts on heritage while developing and executing adaptation and mitigation strategies. Expanding networks of support, fostering collaboration between stakeholders and communities, and encouraging open dialogue are essential steps in designing effective solutions to mitigate the impacts of climate change on maritime cultural heritage.

References

Andrade, C.A. 2008. Cambios recientes del nivel del mar en Colombia, in J.D. Restrepo (ed.) *Deltas de Colombia: morfodinámica y vulnerabilidad ante el Cambio Global*: 101-121. Fondo Editorial Universidad EAFIT, COLCIENCIAS.

Berger, J. and J. Mohr 2009. *Otra manera de contar*. Barcelona: Gustavo Gili.

ColaboraRtorio Azul, viewed 1 December 2025, <https://colaborartorioazul.weebly.com/>.

García Canclini, N. 1999. Los usos sociales del Patrimonio Cultural. Patrimonio etnológico: nuevo perspectivas de estudio. *Patrimonio Etnológico. Noticias perspectivas de estudio*: 16-33.

Del Cairo, C., G. Caro, L. Rozo-Pinzón, L.V. Báez, D. Acosta, N. Palacios, J. Rodríguez and J. Aldana 2022. *Colaborartorio Azul*. Efectos del Cambio Climático es los Bienes de Interés Cultural de Cartagena de Indias, Colombia. Universidad de Cartagena - Instituto Distrital de Patrimonio Cultural (IPCC) - Fundación Colombia Anfibia.

Del Cairo, C., G. Caro, L. Rozo-Pinzón, C. Riera, and J. Aldana 2023. Propuesta de evaluación y monitoreo Co-participativo acerca de los efectos del cambio climático es El patrimonio cultural Costero de Cartagena de indias, Colombia. Universidad de Cartagena - Instituto Distrital de Patrimonio Cultural (IPCC) - Fundación Colombia Anfibia.

Forero, E., Y. Hernández and C. Zafra 2023. Percepción de cambio climático: metodologías, herramientas y estrategias de adaptación es comunidades locales. *Una revisión. Revista UDCA Actualidad & Divulgación Científica* 17.1: 73-85.

ICCROM, Enfoques centrados en las personas, viewed 1 November 2023, <https://www.iccrom.org/es/section/comunidades-y-patrimonio/enfoques-centrados-en-las-personas>

IPCC, 2022, Climate change 2022: Impacts, adaptation, and vulnerability, IPCC Working Group II contribution to the sixth assessment report, viewed 1 November 2023, <https://www.ipcc.ch/report/ar6/wg2/>

Pabón, J.D. 2003. El cambio climático global y su manifestación en Colombia. *Cuadernos de Geografía* XII.1-2: 111-119.

Pabón J.D. and J.A. Lozano 2005. Aspectos relacionados con las estimaciones globales y regionales del ascenso del nivel del mar y su aplicación a Colombia. *Cuadernos de Geografía* 14: 97-106.

Rangel E.S. and J.E. Montealegre 2003. Análisis de las series del nivel del mar en el Pacífico colombiano y su relación con el cambio climático. *Meteorología Colombiana* 7: 53-66.

Rivera-Collazo, I.C. and M. Declet -Pérez 2017. Contribuciones de la arqueología a la mitigación de riesgos ante el Cambio Climático: lecciones recuperadas de Tibes y de Los Bateyes de Viví , Puerto Rico. *Cuba Arqueológica* X.2: 5-15

Scantlebury, M. 2018 'Captain The Island's Sinking!' - Climate Change And Tourism In Speightstown, Barbados, West Indies. Climate Change, International Institute For Sustainable Development (IISD). <https://www.iiis.org/CDs2008/CD2009SCI/SCI2009/PapersPdf/S363PB.pdf>

Torres, R. R. and M. N. Tsimplis 2013. Sea-level trends and interannual variability in the Caribbean Sea. *Journal of Geophysical Research: Oceans* 118.6: 2934–2947. <https://doi.org/10.1002/jgrc.20229>

UNESCO 2014. Gestión del riesgo de desastres para el patrimonio mundial. Paris: UNESCO. <https://unesdoc.unesco.org/ark:/48223/pf0000228134.locale=es>

7.
Traditional Watercraft of the Aegean Sea: A Resource of Local Knowledge and Sustainability

Katerina Velentza

This paper presents an overview of the three-year project 'Re-imagining the use of traditional watercraft in the Aegean Sea for a sustainable environment and economy', which was carried out at the University of Helsinki with funding granted by the Kone Foundation (*Koneen Säätiö*).

The project explored the use of traditional wooden boats, known as *kaikia*, in the Aegean Sea, in Greece, as a historical resource of local knowledge and an opportunity for reintroducing sustainable ways of living to the society, environment and economy of the Aegean. Through the desk-based research of archival resources and existing literature on uses of these traditional boats in Greece, combined with data collected during a maritime ethnographic fieldwork on the islands of Samos and Patmos, complemented with social science data gathered from residents of the Aegean, this study highlights the potential of maritime heritage to assist in restoring the balance between humans and nature in the dynamic environment of the eastern Mediterranean.

This interdisciplinary study presents the potential of utilising historical, archaeological and ethnographic approaches in assisting communities to build environmental awareness and realise sustainability opportunities in the face of the current climate emergency and environmental deterioration. Overall, the scope of the project highlights the contribution that humanities and heritage research could have to building community resilience for the global challenges humanity is facing at present. Recording and learning about the past can help to build a better future.

Introduction

The unprecedented human-induced climate emergency and environmental deterioration that humanity is facing at present has created an urgent need to act to avoid an unparalleled disaster (e.g., Anderson and Bows 2011; Crumley 2021; Erlandson and Braje 2013: 1–3; IPCC 2019, 2021, 2022; Van de Noort 2013: 6–10). Beyond the measures required by governments, stakeholders and businesses, defined in various international agreements (e.g., Paris Agreement 2015), the promotion of sustainable lifestyles at an individual and community level has been identified as equally important for stopping the over-exploitation of the planet's resources restoring the balance between humans and natural ecosystems and building climate resilience (e.g., Archer 2016; Cook 2019; McDonagh *et al.* 2023; Welch and Southerton 2019). However, misinformation, indifference, scepticism and daily-life hardships, including various financial and social crises, brought by political, social and economic circumstances of the 21st century, have been hindering the process.

Recent research initiatives have shown the significance of education, knowledge exchange, community engagement and use of participatory methods to communicate and explain the urgent need for change in the present climate and environmental crisis (e.g., Altinay 2017; Smith *et al.* 2024; Welch 2017). Moreover, local, traditional and indigenous knowledge rooted in ancestral practices and cultural understandings of the natural environment (e.g., IPCC 2022; Virtanen *et al.* 2020), including the human and environmental past (e.g., Crumley 2021; Holthaus 2008; Sabaté and Soret 2014), which take into account patterns of social behaviour and human psychology (e.g., Axsen *et al.* 2012; Lorenzen 2012; Oskamp 2000) and community needs (e.g., Archer *et al.* 2014; Dodman and Mitlin 2011) could help people take action against environmental degradation as well as adjust and improve their ways of living.

This paper builds upon findings from the project 'Re-imagining the use of traditional watercraft in the Aegean Sea for a sustainable environment and economy' ('Re-imagining traditional watercraft' hereafter) and demonstrates how traditional boats, and various traditional social practices surrounding their construction and use, could improve the human-environment interactions in the Aegean Sea, in Greece, for the present and the future.

The paper starts with a brief description of the social and historical background of using traditional boats in the Aegean Sea region in post-medieval periods, the cultural significance of this local watercraft (as tangible and intangible heritage) and the changes that have occurred since the second half of the 20th century. Subsequently, the aims and methodology of the project are explained, followed by the results of the research. Finally, the conclusion outlines the significance of the project results and their potential to influence sustainability initiatives in Greece and beyond. Overall, this study attempts to expound the social and environmental significance of the local knowledge surrounding traditional watercraft for the Aegean Sea communities, along with the potential benefits of the better preservation, safeguarding and re-introduction of such tangible and intangible heritage in the Greek seas.

Figure 7.1: Kaiki at Linaria, Skyros, in 1928. Copyright: Georgios Vafiadakis, Digital Collections of the Hellenic Literary and Historical Archive (ELIA), CC BY 4.0.

Socio-historical background

The traditional wooden boats (Figure 7.1), known as *kaikia* (*καΐκια* in Greek, *καΐκι* in singular; seen also in literature as caique e.g., Smylie 2009: 125), were established technologically in the Aegean Sea (Figure 7.2) roughly in the 18th century (Damianidis 1998: 13–30; Intangible Cultural Heritage of Greece n.d.). Although there is no secure way to prove the direct link between ancient shipbuilding practices and the construction of post-medieval and contemporary traditional boats, it is thought that the *kaikia* are the result of several centuries of human interactions with the surrounding natural environment and an evolution of shipbuilding techniques that the local communities developed in response to their environmental and social needs (Basch 1972: 10; Gillmer 1973: 124–129; Throckmorton 1964). The attachment of the Aegean residents to the sea, the weather and navigation conditions in that maritime setting, as well as the availability of particular material resources, such as certain species of timber, heavily influenced the development of specific types of wooden watercraft, as analysed previously by Damianidis (1998: 41–100).

Figure 7.2: Map of the Aegean Sea, in the eastern Mediterranean, between Greece and Türkiye. Copyright: Katerina Velentza.

Up until the second half of the 20th century, these wooden boats were a defining element of the Aegean populations and their maritime identities, present in all economic and social activities. They were constructed empirically with local timber, propelled by sails, oars (Figures 7.3-7.4) or, since the middle of the 20th century, a combination of those and mechanical inboard engines (Figure 7.5). They were used for the transport of people and goods (Figures 7.6-7.7) and fishing (Figure 7.8), while also having significant cultural and religious associations (Figure 7.9) for coastal and island communities (e.g., Damianidis 2023; Kizos *et al.* 2007; Myres 1941). These boats, as local tangible heritage assets, and the practices of building and using them, as traditional crafts and intangible heritage, have been expressions of local knowledge of the Aegean Sea communities, which is defined by the IPCC (2022: 2914) as 'the understandings and skills developed by individuals and populations specific to the places they live that inform decision making about fundamental aspects of life'.

Figure 7.3: Postcard with photograph of kaikia exiting the port of Kalamata, 1912. Copyright: Digital Collections of the Hellenic Literary and Historical Archive (ELIA), CC BY 4.0.

Figure 7.4: Fishing boat at Stylida, 20th century. Copyright: Digital Collections of the Hellenic Literary and Historical Archive (ELIA), CC BY 4.0.

Figure 7.5: Picture of contemporary fishing kaiki with inboard engine at Ormos Marathokampou, Samos, Greece, in 2022. Copyright: Katerina Velentza.

Figure 7.6: Passengers on a kaiki off the island of Skyros, in 1940. Copyright: Digital Collections of the Hellenic Literary and Historical Archive (ELIA), CC BY 4.0.

Figure 7.7: People on a kaiki. Copyright: United Photojournalists Agency Athens, Pesmazoglou Arcade; from the Digital Collections of the Hellenic Literary and Historical Archive (ELIA), CC BY 4.0.

Figure 7.8: Fishermen loading their nets on a kaiki, 1930. Copyright: Frédéric Boissonnas, Digital Collections of the Hellenic Literary and Historical Archive (ELIA), CC BY 4.0.

Figure 7.9: Greek Orthodox Epiphany celebration (6th of January). Devouts on boats watch the sanctificantion of the waters at Thessaloniki, 1919. Copyright: Digital Collections of the Hellenic Literary and Historical Archive (ELIA), CC BY 4.0.

Over the course of the 20th century, due to the rapid modernisation of Mediterranean societies, the Aegean Sea populations were encouraged to shift away from traditional practices and ways of living (Nikolakakis 2017). As part of this development, there was a swift reduction in the need for traditional wooden boats (Damianidis and Zivas 1986; Gillmer 1973). This concurred with a transition to metal or fibreglass motorboats for fishing and to larger ships that took over the transport of passengers and goods in the Aegean. These changes happened gradually (Figure 7.10), but, by the end of the century, resulted in significant losses of traditional maritime jobs, changes in the cultural values, understandings and significance of the maritime landscape for the Aegean populations, the development of touristic rather than maritime economies with seasonal and precarious employment opportunities and the degradation of the marine environment from the extensive use of large-scale polluting fisheries, as well as cargo and tourist ships (Kizos *et al.* 2007). Additionally, since the coastal and island populations started to interact with the sea less in their daily lives, they experienced gradual loss of local knowledge related to living in and with the local marine, insular or coastal environment. Traditional ways of maritime navigation, fishing, knowledge and habits for co-existing with the natural ecosystems were slowly phased out from the lives of the Aegean residents.

Figure 7.10: Postcard with photograph of the port of Kavala, 1972. Traditional sailing kaikia, kaikia with inboard engines, and modern ferries visible. Copyright: Public Benefit Organisation of Kavala 'Dimofelia', CC BY-NC-SA 4.0.

Currently, Greece is experiencing an even sharper decline in wooden shipbuilding and the use of traditional *kaikia*. One of the biggest threats to the preservation of this tangible and intangible heritage is the destruction of wooden fishing boats (Figures 7.11-7.12), funded by subsidies provided by the Greek government (Maragoudaki 2019: 30-31; Papadimitriou 2019). Especially since 2014, financial compensation has been offered to artisanal fisheries operating traditional boats, if the owners permanently cease their fishing activities, in an attempt to enforce the 'EU Regulation No 508/2014 of the European Parliament and of the Council of 15 May 2014 on the European Maritime and Fisheries Fund' and the subsequent EMFAF Regulation (EU) 2021/1139 that aim at reducing the fishing fleet and the environmental impact of the fisheries of each European state. This is despite explanations from the EU Commissioner for Environment Oceans and Fisheries, Virginijus Sinkevičius, to a query by the President of Hellenic Maritime Museum that there is no legal obligation for the Greek authorities to demand the physical dismantling of any vessel in order to implement the relevant legislation and reduce the impact of fisheries (Damianidis and Anagnostopoulou-Paloubi, personal communication June 21, 2023). Consequently, artisanal fishers with traditional wooden boats have been targeted by the Greek government and offered substantial monetary funds, higher than the actual worth of their boat, in exchange for their fishing licence and the destruction of their fishing vessel. According to the Traditional Boat Association of Greece (n.d.) and Papadimitriou's (2019: 83–84, tables 2–3) research, this has resulted in the destruction of thousands of traditional boats, funded by consecutive government schemes offered since the 1990s. So far, the Greek government has failed to address the issue of deliberately destroying handcrafted traditional boats through the subsidies offered in previous years.

Figure 7.11: Destruction of the kaiki 'Alekos' with government subsidies at the island of Syros, Greece, in 2018. Copyright: Georgia Papadimitriou.

Figure 7.12: Destruction of the kaiki 'Analipsi' with government subsidies at the island of Syros, Greece, in 2018. Copyright: Georgia Papadimitriou.

Aims and theoretical background

The project 'Re-imagining traditional watercraft' had three main aims. Firstly, to create a representative academic record of the community significance and the state of tangible and intangible heritage related to traditional boats in the Aegean Sea in Greece in the 21st century before the assets, local knowledge and related intangible data disappear from lived experiences. Secondly, to assess the potential of utilising local knowledge and maritime heritage to pursue sustainability targets for the Aegean populations. Thirdly, to initiate awareness amongst local communities, stakeholders and heritage professionals of the importance of safeguarding cultural heritage alongside natural heritage, as suggested by the UN Sustainable Development Goal Target 11.4 (UN Sustainable Development Goals n.d.), not simply for historical and cultural reasons, but as an effective means to mend the broken bond between people and the environment in the rapidly deteriorating maritime landscape of the Aegean region.

Given Damianidis' (1991, 1998) meticulous documentation of all aspects of traditional Greek shipbuilding in the 1980s and 1990s, this study did not focus on creating a detailed technical record of how traditional boats are built. On the contrary, this research focused on the interconnections between the Aegean communities and the maritime environment as revealed through the local knowledge surrounding the construction and use of traditional boats. Moreover, the project has provided academic insights to assist local organisations, such as the Society of Friends of the Museum of Aegean Boatbuilding and Maritime Crafts (n.d.), and the Archipelago Network (see Chapter 8), in their advocacy for the protection of traditional boats, crafts and ways of living to support sustainable and resilient ways of living in the Aegean Sea.

To address the project aims, the theoretical approach is grounded within the field of maritime ethnography – so chosen for its capacity to record and rescue contemporary maritime cultures and traditional maritime ways of life and watercraft that are rapidly changing or disappearing (Blue 2003: 334; McGrail 2001: 3). This is a well-used and highly effective approach for the comprehension of the rapid loss of wooden boats in the Aegean in recent years due to government-subsidised destruction, as well as various socio-economic and environmental risks posed to the local maritime populations by climate change.

Moreover, traditional boats of the Aegean Sea, as tangible cultural heritage assets, and the practice of wooden shipbuilding, as a traditional craft and intangible heritage, have been considered in this project as parts of the maritime cultural landscape of the region. This theory, as defined clearly by Westerdahl (1992; 2011) and applied by many other scholars (e.g., Tuddenham 2010), explores a unified view of the maritime sphere, combining land and water realms, and follows the intricate, dynamic, and constantly changing entanglements of people and their culture with the sea. Additionally, aspects of intangible heritage related to the building and use of traditional boats are considered expressions of the Aegean people's local knowledge. Local knowledge (seen also in scholarship as traditional (ecological) or indigenous knowledge) as a theoretical concept expresses deep knowledge systems developed by societies in the long-term through interaction with their surrounding natural environment. The local knowledge of creating and using wooden boats in that sea region had been a defining element of the maritime cultural landscape of the island and coastal populations.

By adopting the theoretical aspects of local knowledge, along with the maritime cultural landscape and Adams' (2001: 300–303, fig. 1) theory on the interrelated constraints surrounding the form, structural characteristics, appearance and use of watercraft for the group of people constructing them, this project has been interpreting the decline in traditional boat use and the changes that occurred in the shipbuilding customs in the late 20th and early 21st centuries within the social, economic, and environmental circumstances of that period. Therefore, it has been possible to initiate an applicable assessment on the significance, vulnerability, and longevity of traditional wooden shipbuilding and traditional boat use. Finally, this paper follows ideas on how maritime archaeology as well as tangible and intangible heritage can be utilised to promote environmental awareness, sustainability and climate resilience argued by the author and other scholars of the field (e.g., Holly *et al.* 2022; Trakadas 2022; Velentza 2023).

Figure 7.13: Member of the photographer's family travelling on a kaiki, c.1900. Copyright: Michail Veloudios, Digital Collections of the Hellenic Literary and Historical Archive (ELIA), CC BY 4.0.

Figure 7.14: Fishermen on a kaiki, off the island of Psara, 1961. Copyright: Dimitris Papadimos, Digital Collections of the Hellenic Literary and Historical Archive (ELIA), CC BY 4.0.

Methodology

The first methodological approach used for data collection in the project was desk-based background research on existing information surrounding the maritime and shipbuilding traditions of the Aegean (e.g., Damianidis and Zivas 1986; Damianidis 1991, 1998; Dimitriadis and Landros 2014; Papadimitriou 2019). As part of this, online archives with historical pictures were also explored, for example the photography collection of the American photographer Robert A. McCabe (n.d.), the archives available through the National Aggregator of Digital Cultural Content in Greece (SearchCulture.gr) and the digital collections of the American School of Classical Studies at Athens (n.d.). In these records, a search was carried out for digitised historical pictures with the terms '*βάρκα*' (boat), '*καΐκι*' (*kaiki*), '*καραβομαραγκός*' (boatbuilder), '*ταρσανάς*' / '*καρνάγιο*' (shipyard) and '*ψαράς*'/'*ψαράδες*' (fisherman/ fishermen). This research revealed a vast amount of archival information recording the breadth of use of traditional boats in the daily lives of Aegean residents (Figure 7.13), as well as their significance for people's survival and livelihoods in that maritime landscape (Figure 7.14). However, due to the inconsistency of these digitised photographic records and lack of categorisation and archiving through maritime or nautical criteria, it was not possible to use the data to conduct a quantitative analysis or a deeper interpretation of these sources. Therefore, the historical photographs have been used as indicative records of traditional boat uses during the 20th century.

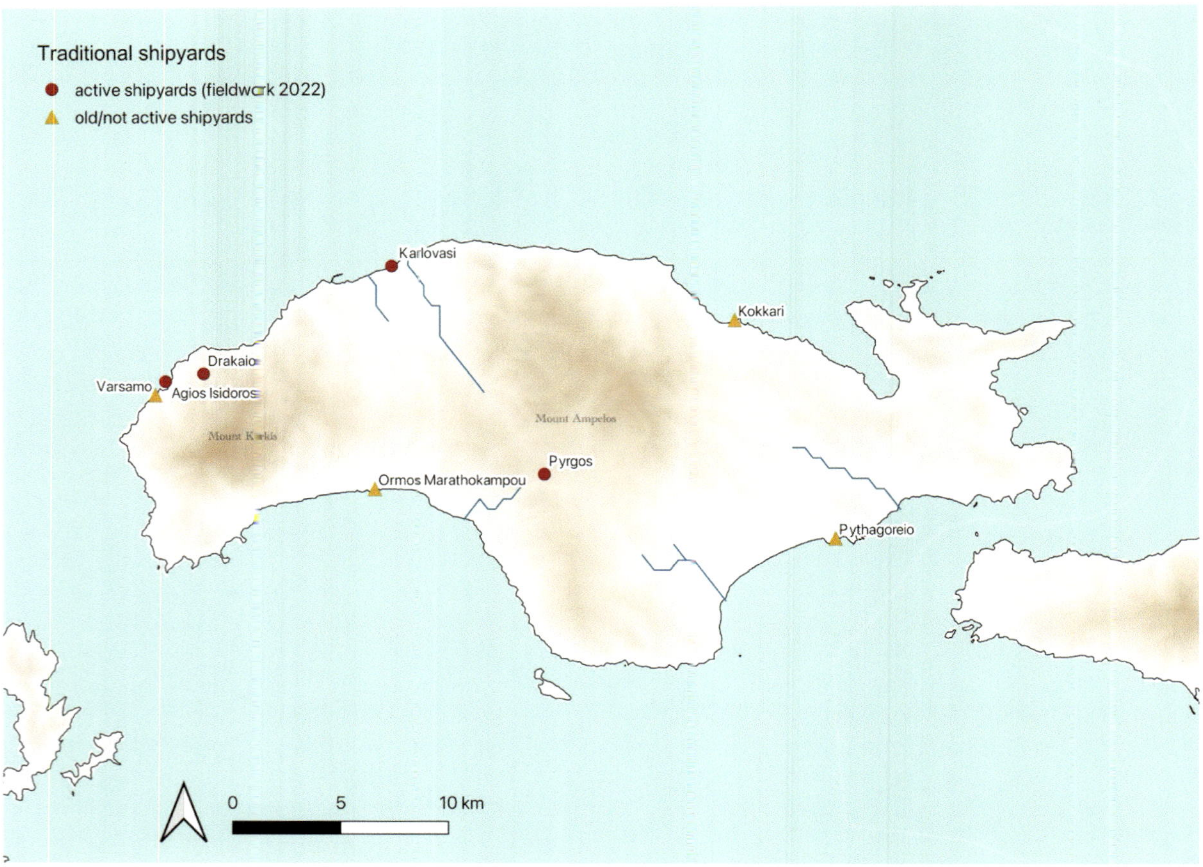

Figure 7.15: Map of active and old/not active shipyards at Samos. Copyright: Katerina Velentza.

The main methodological approach of this project was the maritime ethnographic fieldwork which took place on the islands of Samos and Patmos in September 2022. During the fieldwork, two methods were used. Firstly, photographic documentation capturing aspects of the material record observed in four operating traditional shipyards of Samos (Figure 7.15) and one shipyard of Patmos. Incidental traditional wooden boats or remains of them and pieces of shipbuilding equipment were also recorded with photography on an *ad hoc* basis (Figure 7.16). With this approach it was possible to document aspects of the tangible material assets and the characteristics of traditional shipyards, some of the least often investigated industrial archaeological sites according to Moser (2011: 835). The second method of data collection was semi-structured interviews with shipbuilders, which were conducted in Greek and recorded through note taking and digital audio recording. The interviews took place in the four shipyards of Samos, and the one in Patmos, during working hours. The participation of the shipbuilders in those interviews was voluntary and based on informed consent following the 2019 guidelines on 'The ethical principles of research with human participants and ethical review in the human sciences in Finland' of the Finnish National Board on Research Integrity TENK. This documentation provided significant oral testimonies from the last few carriers of local knowledge related to the construction of traditional watercraft, complementing previous and current local research by Papadimitriou (2019) and the Archipelago Network, presented also in Chapter 8 of this volume.

Figure 7.16: Traditional oared kaiki on the beach at Ormos Marathokampou, Samos, Greece, in 2022. Copyright: Katerina Velentza.

Following the fieldwork, the island of Samos was selected as the focal case study of the project due to the number of traditional shipyards preserved there, as well as the active rescue work of the Society of Friends of the Museum of Aegean Boatbuilding and Maritime Crafts (n.d.) on the island (Velentza 2024). Given the depth of the recordings of the tangible heritage and visible elements of the traditional shipyard infrastructure still preserved on Samos (Figure 7.17) along with the oral histories, living memories and the personal perspectives of individual shipbuilders, this case study enabled the consideration of traditional shipbuilding and its decline within the larger socio-economic context and the lived experiences of the maritime populations of the Aegean (Ransley 2011: 890).

Figure 7.17: Abandoned bandshaw at an old shipyard site at Ormos Marathokampou, Samos, Greece, in 2022. Copyright: Katerina Velentza.

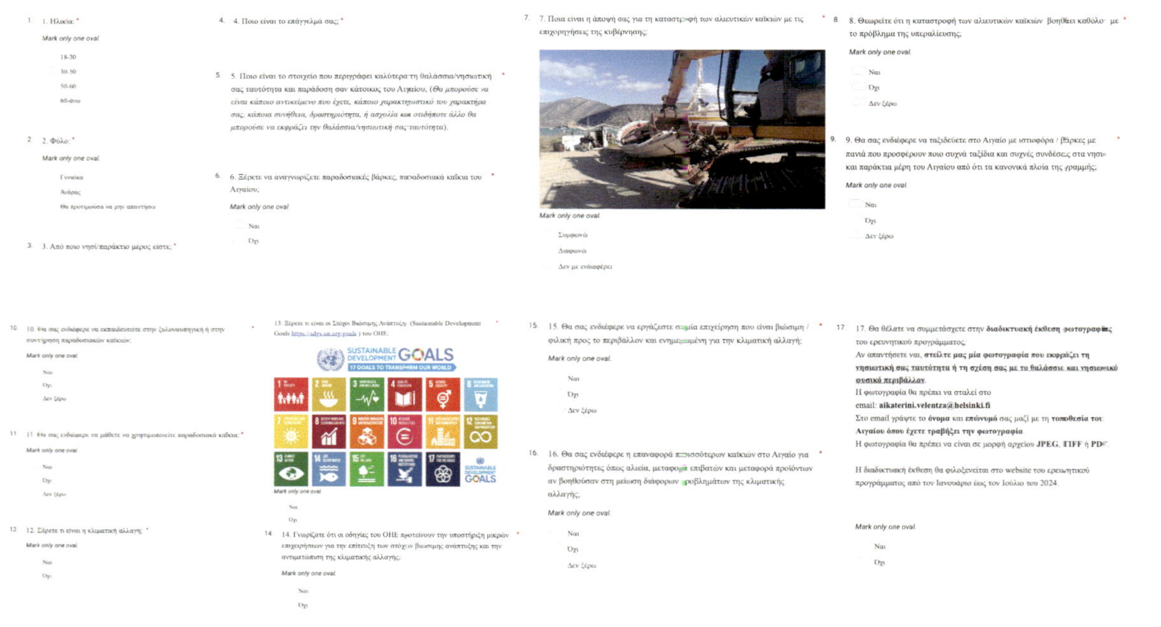

Figure 7.18: Questionnaire for residents of the Aegean used for project 'Re-imagining the use of traditional watercraft'. Copyright: Katerina Velentza.

Vangelis D. Dardanos, Chalkida.

Maria Karla, Fourni Korseon.

Figure 7.19: Screenshot from the online exhibition 'Maritime identities of the Aegean', available at: https://traditionalwatercraftaegeansea.wordpress.com/online-exhibition/. Copyright of selected pictures: Vangelis D. Dardanos and Maria Karla.

Finally, the research project was complemented by social science data collected through an online questionnaire, which was distributed through mailing lists, social media and local contacts in Greece. The questionnaire was directed to residents of Aegean islands and coastal sites in Greece. It consisted of 17 questions (Figure 7.18) enquiring on topics related to traditional boats, shipbuilding, maritime identity, climate change and sustainable development. It was completed anonymously by 27 participants, and it provides an indicative sample of the local attitudes towards maritime traditions and heritage related to the *kaikia* and their potential for the future. Moreover, through this survey, participants were asked to submit photographs that express their maritime identity, compiling thus the online exhibition of the project *Maritime identities of the Aegean* available also on the project website. Despite the small number of the survey participants, the frequency of submitted pictures featuring the sea or traditional boats (Figure 7.19) highlights the continuing connection of Aegean people to the dynamic maritime environment of the region, through this watercraft, despite the loss of key local knowledge and intangible heritage in the past few decades.

The use of the above methodologies and sources (archival, ethnographic and survey data) enabled the reconstruction of social and material perspectives related to the use of traditional boats in the Aegean along with the consideration of changing social contexts and relationships with the natural environment over time.

Results

Changing values and social significance

As explained in the socio-historical background above, the wooden boats, constructed empirically with local timber, have been integral parts of the collective identity of the maritime populations of the Aegean Sea for thousands of years. As Damianidis *et al.* (1998) analyse, shipbuilding and sailing enabled Aegean populations to turn the sea into a means of connectivity and mobility rather than a barrier. The particular design, technology and methods of shipbuilding, which were established roughly in the 18th century (Damianidis 1991: 13–30) and still exist today, defined the population's social and economic activities, their practices and their maritime identities. Damianidis (1998 101–107) explains how the morphology of each type of traditional boat, including the shape of the hull and the methods of propulsion, was influenced by the purpose and use of the watercraft. This note matches Adams' (2001: 300–304, fig. 1) observations on the interrelated constrains that determine the form, structural characteristics, appearance and use of a watercraft. To construct and use this watercraft, local populations were required to know details of the natural environment they lived in, and to learn how to navigate it and make their livelihoods from it. Thus, the cultural elements of traditional shipbuilding and boat use, and with their evolution over time based on the social needs of the populations, has been part of the local maritime experience that Westerdahl (1992: 8) calls 'tradition of usage'.

When historical, political and social changes of the 20th century forced populations to move away from traditional ways of living, the social value of traditional wooden boats and shipbuilding changed drastically, affecting the maritime identity of local people. For the island community of Samos, there has been an obvious erosion of the local significance of the shipbuilding craft, as well as fishing with traditional boats. Today, the last shipbuilders of Samos (Figure 7.20), even though the last carriers and custodians of deep knowledge systems involving the empirical understanding of their island and its environment, the sea and the forest, have been spatially isolated and socially downgraded by the local stakeholders. It is noteworthy that shipyards at Agios Isidoros and Pyrgos, and the one in Karlovasi set in a former industrial area of the town, are situated in remote locations (Figure 7.15), away from people rather than within central areas of a village or town, as observed by

Damianidis (1991: 31–32) in the 1980s. With the removal of the shipyards from the interior physical space of towns and villages we see a change in the cultural space and the cultural values of these communities. The remoteness, isolation and social marginalisation of the shipyards demonstrate the loss of social significance of that craft for the local maritime communities as well as a change in their priorities as maritime people. Furthermore, the mountainous setting of the shipyards at Drakaioi and Pyrgos indicate the reduced need for routine maintenance and boat repairs (Damianidis 2023: 123), which would require shipbuilders to work closer to the seafront (Moser 2011: 839).

Figure 7.20: Shipbuilder working at the shipyard of Pyrgos, Samos, in 2022. Copyright: Katerina Velentza.

Some shipbuilders on Samos have received complaints from individual councillors of their local municipality about the destruction of the island landscape with their shipbuilding activities. For example, the shipbuilders at the protected heritage site of Agios Isidoros (Figure 7.21) noted in their interviews that they have been blamed for 'polluting' the cove with their wooden chips and timbers, which have been characterised by municipality councillors as 'rubbish' that turns away tourists, even though there is no pollution or polluting waste from traditional wooden shipbuilding. The testimonies of such attitudes of local government officials show a local change of aesthetics and a tendency to consider tradition as hindering progress, noted also in the research done by Dragouni and Lekakis (2023) elsewhere in Greece.

Figure 7.21: The shipyards of Agios Isidoros at Samos, Greece, in 2022. Copyright: Katerina Velentza.

Moreover, the manpower of shipbuilders has reduced rapidly in the 21st century. Giorgos Kiassos, the shipbuilder at Drakaioi remembered: 'Back when I went to Agios Isidoros [to learn the craft of shipbuilding], each shipbuilder had three or four young men learning and there were around five shipyards'. The cove was extremely busy and constituted basically a school for young men to learn traditional shipbuilding, which is not the case anymore (Damianidis 2023: 122).

These changes in the physical locations and social consideration of the traditional shipyards, along with the reduction of shipbuilding workers, the smaller capacity of shipbuilding in each shipyard and the lower demand in traditional boats for small-scale fishing, all due to government-subsidised attempts to reduce professional fishing licences, signify a social and cultural transformation in the region.

Changing relationships with the natural environment

Since the second half of the 20th century and the post-war attempt to bring the Greek state to the standards of other western countries, consecutive Greek governments and local policymakers urged communities to move away from traditional crafts and ways of living to modernise the Aegean, provide economic and social development, and advance the population's quality of life and their economies. Tourism has played a central role in this reform (Alifragkis and Athanassiou 2013; Tsartas 2003). The new socio-economic circumstances that the modern, 'advanced', world imposed in the Aegean have taken a toll on local knowledge and ancestral ways of living by the sea, which bridged nature and culture. This pursuit of modernity at the cost of heritage, tradition and balanced relationship with the natural environment is a common feature of western maritime communities recorded by other scholars, such as Nadel-Klein (2003) working with fishing communities at the Scottish coast. The decline in the use, construction, and appreciation of traditional wooden boats has affected the social and ideological associations of the Aegean *kaikia* and created a shift in the maritime cultural landscape of the region. This shift is also reflected in changes of the maritime identities of the Aegean populations. Unlike the close familiarity of Aegean residents with traditional boats, navigation and sailing conditions at sea seen in the early 20th century (Figures 7.22-7.23), these same populations of the Aegean have become very unfamiliar with core aspects of maritime life in their region. As the ethnographic research and questionnaire of the project highlighted, Aegean residents are rarely familiar with sailing or operating traditional boats and therefore unaware of maritime navigation or climate- and weather-related patterns of their surrounding environment.

Figure 7.22: A group of friends in a trawler, c.1910. Copyright: Digital Collections of the Hellenic Literary and Historical Archive (ELIA), CC BY 4.0.

In parallel to the gradual decline in the number of traditional boats supporting the Aegean islands for transport and subsistence, through the course of the 20th century and peaking in the 21st century the coastal and island communities have developed increasingly more continental views and ways of living (Kizos 2007). The dependence of the residents on the main line ferries or short flights, the only possibilities for transportation links to other islands, Athens and mainland Greece, as well as the promotion of exploitative tourism as the main, and sometimes only, source of income for island populations (Soilemezidou and Orfanidi 2022) has worn down the connection of the locals with the marine environment (Hunter 2002). Most of the residents of Samos interviewed in 2022 described the extreme isolation they experience on the island especially during the winter months with lack of vital resources and job opportunities. Sailing at sea in smaller boats with frequent connections is not a common habit as it used to be. The sea is seen as a tourist attraction and therefore a financial resource to be exploited, but otherwise an obstacle or a vice that the local communities must overcome during their lives.

Figure 7.23: Kaiki with passengers at Platys Gyalos, on the island of Euboea/Evia, c.1920s. Copyright: Digital Collections of the Hellenic Literary and Historical Archive (ELIA), CC BY 4.0.

Figure 7.24: Small lofting line plan known as 'sala' (σάλα in Greek) at the shipyard of Pyrgos. Copyright: Katerina Velentza.

Similarly to this disconnection from the sea, people have been losing touch with their land, too. One striking example is the change to the forests of Samos, noted by local shipbuilders. The work of shipbuilders has always been empirical, based on experiential understandings of the natural environment transmitted from generation to generation, their lived experience, and constant observations of the conditions of the sea and the trees, rather than strict calculations and plans used in modern shipbuilding (Figures 7.24). The shipbuilders have not been simply carving the timbers to make a boat, but, traditionally, they have been the ones selecting and cutting the timbers in a way that managed and maintained the local forests. The shipbuilders interviewed in Samos explained how they construct each boat by hand, carefully shaping and building each unique piece. They consider this work more similar to the process of a sculptor rather than a carpenter because they have the knowledge of how timbers behave and can match the appropriate parts of a tree to the correct parts of each boat. Giorgos Kiassos at Drakaioi noted that: 'every piece of wood has to be the appropriate for where it will go on the vessel' and every boat constructed is unique that cannot be exactly replicated. Because of this deep understanding and strong link that the shipbuilders develop with their raw material, the local Samian timbers, and the watercraft they produce, some consider each boat as a person, with whom they can converse. Giorgos Petrou of the shipyard in Pyrgos compared the construction of a boat to 'a human being born' and he explained how he feels the 'camaraderie' between him and each boat that he constructs. In Samos this traditional process of selecting timbers from their local forests and turning them into wooden boats was so significant that it was decided to protect these natural environments from any other possible exploitation. For example, in the 19th century the local government of Samos established law no. 933, which prohibited resin extraction from any pine tree of the island to protect the Samian forests and preserve them for shipbuilding (Damianidis 1991: 248; Kakaras 2014). This legislation meant that unlike other Aegean islands, like Lesvos and Rhodes, that extracted and sold extensively resin from their forests during that period of history, the local administration of Samos tried to stop exploitative practices to maintain the health and quality of the local forests. This measure, and the tradition it was based on, meant that Samos still maintains its woodland extensively, which is not the case in other islands of the eastern Aegean. The sustainability and importance of Samian forests is also attested during the late 20th century, when in the 1980s Damianidis (1991, 1998: 16–18) reports that most traditional shipyards operating all around Greece preferred to use Samian timber for shipbuilding because it was considered the most appropriate for creating durable wooden vessels for the Aegean Sea conditions due to its health and high content of resin (Damianidis 1991: 232, 243, tables 37–39; 1998:169–178, tables 19–20, fig. 211–213, 2014b; Damianidis and Zivas 1986: 74).

In recent years, though, despite Samos being still well vegetated with pine (*Pinus brutia* and *Pinus nigra*), cypress (*Cupressus sempervirens*) and oak (*Quercus coccifera*), there has been extensive damage to the forests from the frequent wildfires and the disruption to the traditional management practices with modern governing models (Palaiologou and Kalabokidis 2022). Especially at the centre of the island at Mount Ampelos, around the shipyard of Pyrgos, the trees are considered 'sick'. The shipbuilder Giorgos Petrou reported that he used to retrieve lumber for shipbuilding in the forest around his shipyard, but that it all was burnt during the wildfires of the early 2000s. He said: 'We have a problem with the timbers, the timbers are sick, they are not how they used to be'. Because of that damage, now, he can only get the necessary timber from the forest around the village of Idroussa, further north at the other side of Mount Ampelos. On top of the problems with the forest conditions, there are significant issues with timber supply and 'delays with cutting trees because of the poor state of the local forestry office, which is understaffed, underfunded and unable to monitor and manage the forests the way local communities used to do in the past'.

This empirical understanding of the landscape and the natural elements that people of the Aegean had, is part of what Westerdahl (1992: 5–6) called 'cognitive perspective of local tradition', a core part of the maritime cultural landscape theory. These deep knowledge systems, of which shipbuilding has been part, and their influence on local decision making in the past, created sustainable ways of living that could be supported in the long term. However, most of these traditions and knowledge systems are nowadays barely surviving or are completely gone.

Pursuing sustainability through tangible and intangible maritime heritage

The disconnect between maritime people and their surrounding natural environment, with the loss of core heritage and local knowledge described above, are particularly concerning in the context of the current climate crisis and environmental deterioration. The Aegean region has been severely affected by the global rising temperatures, sea level rise, extreme storminess, flooding, and the decline of the marine and island ecosystems. These effects have strained people's lives, while tourism, currently promoted by the government as the 'heavy industry', the central pillar, and main financial resource of the Greek islands and coastal settings (Enterprise Greece, n.d.; Greek Tourism Confederation, n.d.), is draining any remaining local resources. Decades of focus on the exclusively touristic economies of islands and coastal places have not improved the economic wellbeing of the communities for the long term (Tsartas 2003) but have resulted in distancing people from their traditional and sustainable ways of living. The complaints of Samos councillors over the 'pollution' of the Agios Isidoros cove by the shipbuilding activities reported by Vangelis Manoliadis in 2022 and mentioned above is an indicative example of how the local authorities prefer exploiting the natural landscape for touristic purposes rather than historical and traditional activities. Despite this, the seasonality of the tourist industry (Kizos 2007; Zacharatos *et al.* 2014), the environmental deterioration that comes with the thousands of visitors, and the instability of the sector as illustrated with the crash of tourist economies worldwide due to the ban in travelling during the COVID-19 pandemic (e.g., Donaire *et al.* 2021; Duro *et al.* 2021; Lebrun *et al.* 2022), along with the change of tourist behaviours due to climate-induced risks, such as wildfires, heatwaves and flooding (e.g., Kovačić *et al.* 2020), are leaving the Aegean populations in precarious circumstances with no support for the future. The community's lack of understanding of and connection to the environment they live in, the lack of traditional tools and lifestyles, along with the attempt to carry out continental ways of living in island and coastal settings, is creating big disadvantages for these populations. For example, the high prices and the scarcity of ferry lines connecting the marginal islands of Greece, described in the reports by Aegean Cargo Sailing (n.d.) and also analysed in articles by Giannopoulos and Boulougaris (1995), as well as Hatziioannidu and Polydoropoulou (2022), suggest that the current infrastructure for mobility and transport in the Aegean islands cannot fulfil the needs of the local maritime communities.

Data produced through the research of 'Re-imagining traditional watercraft' suggests that the better preservation and protection of tangible and intangible elements of maritime cultural heritage relating to construction and use of traditional wooden boats in the Aegean Sea could be a simple way of maintaining, safeguarding and even re-introducing key elements of local knowledge to the Aegean populations. This could help them to rethink their links to the natural environment. For example, the energy efficiency of traditional wooden boats, as explained by the shipbuilders of Samos (Velentza 2024) and suggested previously by Rassia and Tsikis (2020), as well as their durability and excellent sailing capacity present a unique opportunity to lower the carbon footprint of transport in the Aegean, while also interconnecting the Greek islands with shorter and more frequent transport links, as it used to be in the past. Actually, 100% of the responders in the questionnaire of the project answered that they would endorse the reinstatement and the reintroduction of traditional *kaikia* in the Aegean for a variety of maritime activities including fishing and transport of products and passengers, especially if this action helped to tackle various climate change challenges (Figure 7.25). Also, 81.5% of the respondents expressed interest in travelling with traditional *kaikia* that offered frequent connections in the Aegean (Figure 7.26). Furthermore, the results of the same survey showed that 85.2% of the responders would be interested in learning to use traditional wooden boats (Figure 7.27) and 81.5% would be interested in learning how to build and/or repair traditional *kaikia*.

Even though the participants of the questionnaire are only a very small sample of the overall Aegean population, the striking unanimous support of the potential reintroduction of traditional boats along with the significant interest in learning practical aspects related to traditional boats, highlight that the local population would easily re-embrace some of their past traditions and ways of living to build climate resilience and sustainability. Controlled tourism with the sustainable utilisation of local resources (natural and cultural) could be part of this new reality (Dragouni 2017). As reported by the fishermen interviewed *ad hoc* at Ormos Marathokampou, and as recorded in previous academic studies (e.g., Belias *et al.* 2018), travellers have been increasingly turning away from mass tourism and have grown interest in alternative tourism, where they can experience the local lifestyles in quiet, tranquil, and quality settings.

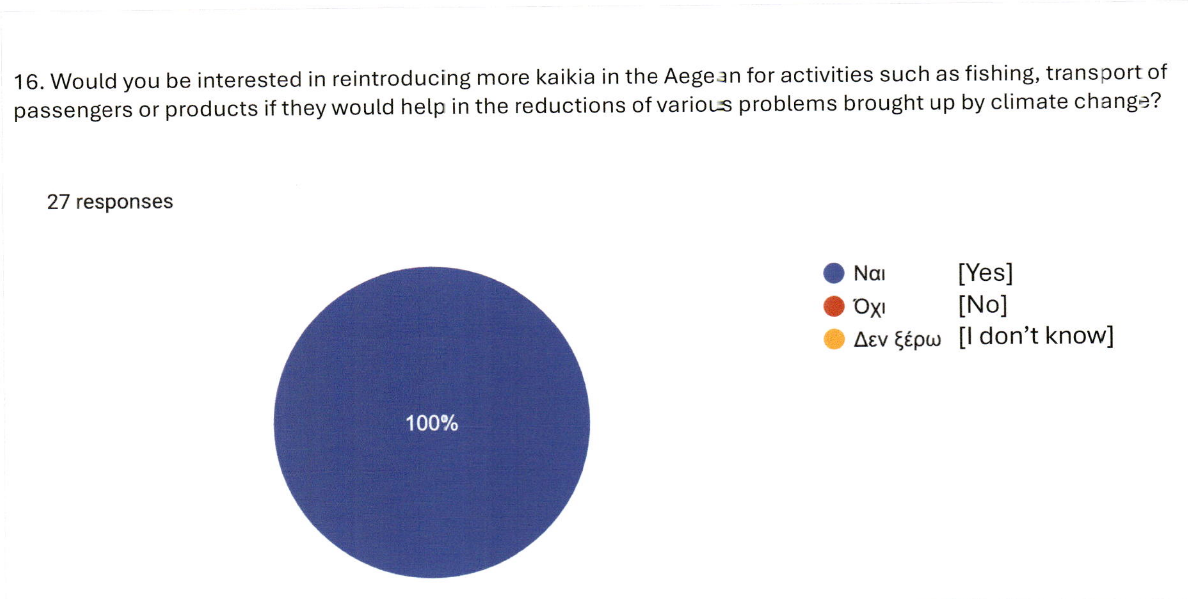

Figure 7.25: Answers to questionnaire query 16: 'Would you be interested in reintroducing more kaikia in the Aegean for activities such as fishing, transport of passengers or products if they would help in the reductions of various problems brought up by climate change?'. Copyright: Katerina Velentza.

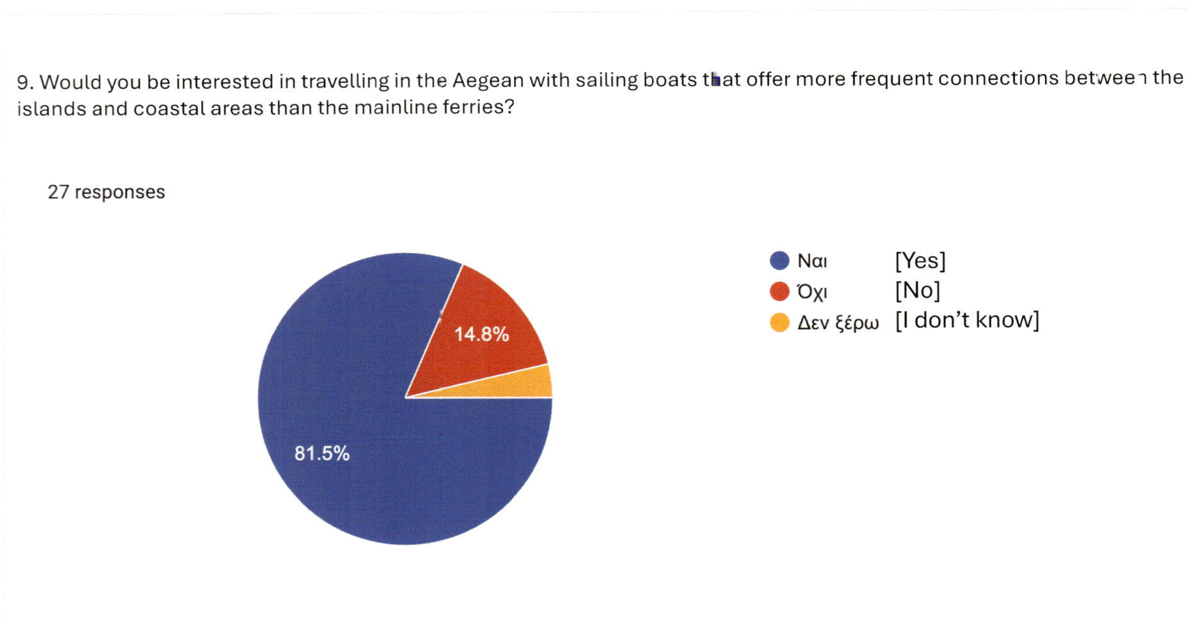

Figure 7.26: Answers to questionnaire query 9: 'Would you be interested in travelling in the Aegean with sailing boats that offer more frequent connections between the islands and coastal areas than the mainline ferries?'. Copyright: Katerina Velentza.

Spending holidays on an agrotourist unit, watching or even helping shipbuilders build traditional boats, sailing between the islands with low or no carbon footprint (e.g., Aegean Cargo Sailing n.d.) are already popular. This type of 'slower' tourism would allow local Aegean communities to develop their economies in a way that will not disrupt their natural resources or hinder their heritage, traditions and local knowledge. Similarly, the ban of large-scale fishing operations, in favour of the preservation of artisanal fisheries using traditional boats, combined with outreach initiatives to inform fishermen about updated regulations and scientifically-informed understandings of the environmental deterioration that the region is experiencing, would allow local maritime communities to be in charge of safeguarding and managing their natural environment and their economies for their own sustainability rather than for country-wide economic development goals within the context of the competitive capitalist markets of the 21st century. These actions could help locals to take charge of their marine citizenship as recommended by the Challenge 10 of the United Nations' Decade of Ocean Science for Sustainable Development 2021-2030 (e.g., Buchan *et al.* 2024).

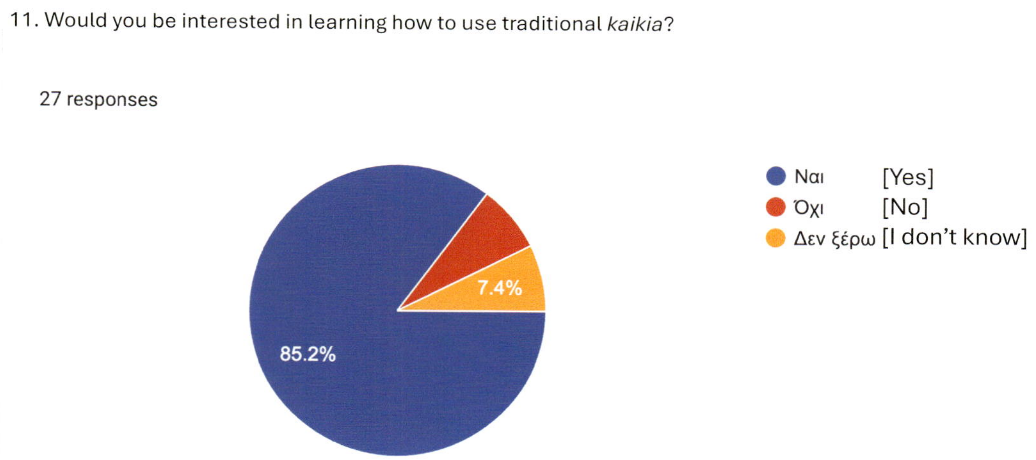

Figure 7.27: Answers to questionnaire query 11: 'Would you be interested in learning how to use traditional kaikia?'. Copyright: Katerina Velentza.

Conclusion

Advisory reports on climate action, sustainability, adaptation and climate resilience (e.g., IPCC 2022) and interdisciplinary academic studies have stated the significance of local knowledge, heritage and culture for helping people adapt and become resilient in the face of climate change and environmental deterioration in the present and the future (e.g., Holly *et al.* 2022; McDonagh *et al.* 2023; Trakadas 2021: 5–7; Velentza 2023: 368–369). Even Westerdahl (1992: 6), writing in the late 20th century prior to the holistic comprehension of the climate emergency, had identified that 'survivals are the product of adaptation' for human cultures and the understanding of the natural landscape (water and land) can play a core part in achieving that. For the Aegean Sea region, in the critical moment humanity is experiencing, there is the potential to preserve and reintroduce traditional watercraft as an easy-to-adopt measure towards sustainability. The deep knowledge systems and sustainable ways of living of the past that traditional boats relate to could benefit immensely the local populations along with their economies, culture, history and future.

'Re-imagining traditional watercraft' has presented a snapshot assessment of the significance, vulnerability and longevity of traditional boats for the lives of the Aegean people. Better preservation and reintroduction of this watercraft would help people reconnect and embrace unique local environments and traditions, promoting, thus, local social and environmental sustainability. From the project data it becomes also obvious that local people have an interest in taking charge of their marine citizenship and adapting their ways of living to mitigate environmental deterioration through culturally aware and tradition-sensitive adjustments. Safeguarding and re-introducing maritime heritage and local knowledge could play a key role in that process because it would provide a familiar background and appeal to elements of the population's past identity and culture. Therefore, maritime communities in the Aegean could have more tools to assist them in adapting to the new reality posed by climate change and building resilience for the present and future. This change, though, would require urgent and decisive action by the government, policymakers and local stakeholders, who largely oppose moving away from modern economic standards and pursuit of growth through tourism. The last decades of financial crisis in Greece have shown that it would be precarious and senseless to consider tourism as a sustainable livelihood and secure financial resource for the future of the Aegean populations. By bringing back traditional watercraft in the centre of the lives of the Aegean maritime communities, people could have an opportunity to reconnect with the natural environment they live in, restore their maritime activities, and reinstate older sustainable practices that have been hindered in the last few decades.

References

Adams, J. 2001. Ships and boats as archaeological source material. *World Archaeology* 32.3: 292–310. <https://doi.org/10.1080/00438240120048644>

Aegean Cargo Sailing, viewed 15 September 2024, <https://aegeancargosailing.org/>.

Alifragkis, S., and E. Athanassiou 2013. Educating Greece in modernity: post-war tourism and western politics. *The Journal of Architecture* 18.5: 699–720. <https://doi.org/10.1080/13602365.2013.838285>

Altinay, Z. 2017. Visual Communication of Climate Change: Local Framing and Place Attachment. *Coastal Management* 45.4: 293–309. <https://doi.org/10.1080/08920753.2017.1327344>

American School of Classical Studies at Athens, Digital Collections, viewed 15 September 2024, <https://www3.ascsa.edu.gr/index-archives.html>.

Anderson, K. and A. Bows 2011. Beyond 'dangerous' climate change: emission scenarios for a new world. *Philosophical Transactions of the Royal Society* 369: 20–44.

Archer, D., F. Almansi, M. DiGregorio, D. Roberts, D. Sharma and D. Syam. 2014. Moving towards inclusive urban adaptation: approaches to integrating community-based adaptation to climate change at city and national scale. *Climate and Development* 6.4: 345–356. <https://doi.org/10.1080/17565529.2014.918868>

Archer, D. 2016. Building urban climate resilience through community-driven approaches to development: Experiences from Asia. *International Journal of Climate Change Strategies and Management* 8.5: 654–669. <https://doi.org/10.1108/IJCCSM-03-2014-0035>

Archipelago Network, viewed 15 September 2024, <https://archipelagonetwork.org/>.

Axsen, J., J. TyreeHageman and A. Lentz 2012. Lifestyle practices and pro-environmental technology. *Ecological Economics* 82.C: 64-74. <https://doi.org/10.1016/j.ecolecon.2012.07.013>

Basch, L. 1972. Ancient wrecks and the archaeology of ships. *International Journal of Nautical Archaeology* 1.1: 1–58. <https://doi.org/10.1111/j.1095-9270.1972.tb00675.x>

Belias, D., E. Velissariou, D. Kyriakou, K. Varsanis, L. Vasiliadis, C. Mantas, L. Sdrolias and A. Koustelios 2018. Tourism consumer behavior and alternative tourism: The case of agrotourism in Greece, in V. Katsoni and K. Velander (eds) *Innovative Approaches to Tourism and Leisure*: 465–478. Cham: Springer. <https://doi.org/10.1007/978-3-319-67603-6_35>

Blue, L. 2003. Maritime ethnography: The reality of analogy, in C. Beltrame (ed.) *Boats, Ships and Shipyards: Proceedings of the Ninth International Symposium on Boat and Ship Archaeology, Venice, 2000*: 334–338. Oxford: Oxbow. <https://doi.org/10.2307/j.ctvh1dsn7>

Buchan, P.M., L.S. Evans, S. Barr and M. Pieraccini 2024. Thalassophilia and marine identity: Drivers of 'thick' marine citizenship. *Journal of Environmental Management* 352: 120111. <https://doi.org/10.1016/j.jenvman.2024.120111>

Cook, J. 2019. *Sustainability, Human Well-Being, and the Future of Education*. Cham: Palgrave Macmillan.

Crumley, C.L. 2021. Historical Ecology: A Robust Bridge between Archaeology and Ecology. *Sustainability* 13: 8210. <https://doi.org/10.3390/su13158210>

Damianidis, K. 1991. Vernacular Boats and Boatbuilding in Greece, Volumes 1 and 2. Unpublished PhD dissertation, University of St Andrews, Scotland. <http://hdl.handle.net/10023/7116>

Damianidis, K. 1998. *Η Ελληνική Παραδοσιακή Ναυπηγική*. Athens: Cultural Technological Foundation of the Industrial Development Bank of Greece.

Damianidis, K. 2023. Μουσειό Ναυπηγικών και Ναυτικών Τεχνών του Αιγαίου: Η προετοιμασία ενός Μουσείου – Εργαστηρίου για την ξυλοναυπηγική στα χρόνια της οικονομικής κρίσης, in P. Karampampas (ed.) *Άυλη Πολιτιστική Κληρονομιά σε Καιρούς Οικονομικής Κρίσης: Ένταξη στην Αγορά και Ανθεκτικότητα*: 121–126. Athens: Greek Ministry of Culture.

Damianidis, K. and A. Zivas 1986. *Το Τρεχαντήρι στην Ελληνική Ναυπηγική Τέχνη*. Athens: Hellenic Organization of Small and Medium Sized Industries and Handicrafts.

Damianidis, K., L. Mendoni, A. Charoni, V. Polychroni, E. Mpechraki and R. Oikonomou 1998. *Σάμος – Παραδοσιακή Ναυπηγική στο Αιγαίο: Ταρσανάδες και Σκαριά*. Athens: Greek Ministry of the Aegean.

Dimitriadis, V. and C. Landros 2014. *Το Σαμιώτικο Πεύκο και η Παραδοσιακή Υλοτομία στη Σάμο*. Pythagoreio: Pnevmatiko Idryma Samou 'Nikolaos Dimitriou'.

Dodman, D. and D. Mitlin 2011. Challenges for community-based adaptation: Discovering the potential for transformation. *Journal of International Development* 25.5: 640-659. <https://doi.org/10.1002/jid.1772>

Donaire, J.A., N. Galí and R. Camprubi 2021. Empty summer: International tourist behavior in Spain during COVID-19. *Sustainability* 13.8: 4356. <https://doi.org/10.3390/su13084356>

Dragouni, M. 2017. Sustainable Heritage Tourism: Towards a Community-led Approach. Unpublished PhD dissertation, University College London, UK.

Dragouni M. and S. Lekakis 2023. Co-creating the future of heritage in-the-making: Empirical evidence from community deliberation at Naxos Island, Greece. *International Journal of Heritage Studies* 29.4: 294–313. <https://doi.org/10.1080/13527258.2023.2181376>

Duro, J.A., A. Perez-Laborda, J. Turrion-Prats and M. Fernández-Fernández 2021. Covid-19 and tourism vulnerability. *Tourism Management Perspectives* 38: 100819. <https://doi.org/10.1016/j.tmp.2021.100819>

Enterprise Greece, viewed 15 September 2024, <https://www.enterprisegreece.gov.gr/en/invest-in-greece/sectors-for-growth/tourism>

Erlandson, J.M. and T. J. Braje 2013. Archaeology and the Anthropocene. *Anthropocene* 4: 1–7. <https://doi.org/10.1016/j.ancene.2014.05.003>

Giannopoulos, G. A. and G. Boulougaris 1995. Overcoming isolation and the role of transport: The case of the Aegean Islands, in H. Coccossis and P. Nijkamp (eds) *Overcoming Isolation: Advances in Spatial Science*: 254–268. Cham: Springer.

Gillmer, T. 1973. *Working Watercraft: A Survey of the Surviving Local Boats of Europe and America*. London: Patrick Stephens.

Greek Tourism Confederation (SETE), viewed 15 September 2024, <https://sete.gr/>.

Hatziioannidu, F. and A. Polydoropoulou 2022. Transport modelling of the ferry network in the aegean archipelagos and its airline alternatives: Analytics, visualisation, and insights. *Maritime Transport Research* 3: 100076. <https://doi.org/10.1016/j.martra.2022.100076>

Holly, G., A. Rey da Silva, J. Henderson, C. Bita, W. Forsythe, Z. A. Ombe, C. Poonian and H. Roberts 2022. Utilizing Marine Cultural Heritage for the Preservation of Coastal Systems in East Africa. *Journal of Marine Science and Engineering* 10: 693. <https://doi.org/10.3390/jmse10050693>

Holthaus, G. 2008. *Learning Native Wisdom: What Traditional Cultures Teach Us About Subsistence, Sustainability and Spirituality*. Lexington: The University Press of Kentucky.

Hunter, C. 2002. Sustainable tourism and the touristic ecological footprint. *Environment, Development and Sustainability* 4: 7–20. <https://doi.org/10.1023/A:1016336125627>

Intangible Cultural Heritage of Greece, Wooden Shipbuilding, viewed 15 September 2024, <https://ayla.culture.gr/xilonaupigiki_wooden_shipbuilding/>.

IPCC (Intergovernmental Panel on Climate Change) 2019, Climate Change and Land, IPCC special report, viewed 15 September 2024, <https://www.ipcc.ch/srccl/>.

IPCC (Intergovernmental Panel on Climate Change) 2021, Climate Change 2021: The Physical Science Basis, IPCC sixth assessment report, viewed 15 September 2024, <https://www.ipcc.ch/report/sixth-assessment-reportworking-group-i/>.

IPCC (Intergovernmental Panel on Climate Change) 2022, Climate Change 2022: Impacts, Adaptation, and Vulnerability, IPCC Working Group II contribution to the sixth assessment report, viewed 15 September 2024, <https://www.ipcc.ch/report/ar6/wg2/>

Kakaras, I. 2014. Το πεύκο και το κυπαρίσσι της Σάμου: Εργασίες συγκομιδής – Εφαρμογές σε ξύλινες κατασκευές, in V. Dimitriadis and C. Landros (eds) *Το Σαμιώτικο Πεύκο και η Παραδοσιακή Υλοτομία στη Σάμο*: 177–198. Pythagoreio: Pnevmatiko Idryma Samou 'Nikolaos Dimitriou'.

Kizos, T. 2007. Island lifestyles in the Aegean islands, Greece: Heaven in summer, hell in winter?, in H. Palang, H. Sooväli and A. Printsmann (eds) *Seasonal Landscapes*: 127–149. Cham: Springer. <https://doi.org/10. 1007/1-4020-4990-0_6>

Kizos, T., I. Spilanis and M. Koulouri 2007. The Aegean Islands: A paradise lost?: Tourism as a driver for changing landscapes, in B. Pedroli, A. van Doorn and G. de Blust (eds) *Europe's Living Landscapes*: 332–349. Zeist: KNNV Publishing. <https://doi.org/10.1163/9789004278073_021>

Kovačić S., M.C. Mărgărint, R. Ionce and D. Miljković 2020. What are the factors affecting tourist behavior based on the perception of risk? Romanian and Serbian tourists' perspective in the aftermath of the recent floods and wildfires in Greece. *Sustainability* 12.16: 6310. <https://doi.org/10.3390/su12166310>

Lebrun, A.M., R. Corbel and P. Bouchet 2022. Impacts of Covid-19 on travel intention for summer 2020: A trend in proximity tourism mediated by an attitude towards Covid-19. *Service Business* 16: 469–501. <https://doi.org/10.1007/s11628-021-00450-z>

Lorenzen, J. 2012. Going Green: The Process of Lifestyle Change. *Sociological Forum* 27.1: 94–116.

Maragoudaki, E. 2019. The boatbuilder, boat building, and the creation of socialities, in C. Papadopoulou (ed.) *The Culture of Ships and Maritime Narratives*: 19–40. Abingdon: Routledge.

McCabe, R., Photography, viewed 15 March 2025, <https://www.mccabephotos.com/>.

McDonagh, B., E. Brookes, K. Smith, H. Worthen, T.J. Coulthard, G. Hughes, S. Mottram, A. Skinner and J. Chamberlain 2023. Learning histories, participatory methods and creative engagement for climate resilience. *Journal of Historical Geography* 82: 91–97. <https://doi.org/10.1016/j.jhg.2023.09.002>

McGrail, S. 2001. *Boats of the World*. Oxford: Oxford University Press.

Moser, J.D. 2011. Shipyard archaeology, in B. Ford, D.L. Hamilton and A. Catsambis (eds) *The Oxford Handbook of Maritime Archaeology*: 834–855. Oxford: Oxford University Press.

Myres, J.L. 1941. The islands of the Aegean. *The Geographical Journal* 97.3: 137–156. <https://doi.org/10. 2307/1787324>

Nadel-Klein, J. 2003. *Fishing for heritage: Modernity and loss along the Scottish coast*. Abingdon: Routledge.

Nikolakakis, M. 2017. The Colonels on the Beach: Tourism Policy During the Greek Military Dictatorship (1967–1974). *Journal of Modern Greek Studies* 35.2: 425–450. <https://dx.doi.org/10.1353/mgs.2017.0026>

Oskamp, S. 2000. A sustainable future for humanity? How can psychology help?, *American Psychologist* 55.5: 496–508.

Palaiologou, P. and Kalabokidis, K. 2022. Emerging challenges of wildfire risk management in the islands of the Aegean archipelago. *Environmental Sciences Proceedings* 17.1: 49. <https://doi.org/10.3390/enviro nsciproc2022017049>

Papadimitriou, G. 2019. Η καταστροφή των καϊκιών – Το κόστος του Πολιτισμού. Unpublished Master's dissertation, Cross-institutional and cross-departmental programme of postgraduate studies, National and Kapodistrian University of Athens, University of Thessaly and University of Western Attica, Greece.

Paris Agreement, 2015, United Nations, viewed 10 September 2025, <https://www.un.org/en/climatechange/paris-agreement>.

Ransley, J. 2011. Maritime communities and traditions, in A. Catsambis, B. Ford and D.L. Hamilton (eds) *The Oxford Handbook of Maritime Archaeology*: 879–906. Oxford: Oxford University Press.

Rassia, S. and T. Tsikis. 2020. Wooden boats and our 'smart sea energy gene': An evolutionary approach to naval architecture and marine engineering through history, optimization, renewable energy, and sustainability. *SN Operations Research Forum* 1: 15. <https://doi.org/10.1007/s43069-020-00018-z>

Re-imagining the Use of Traditional Watercraft in the Aegean Sea for a Sustainable Environment and Economy, project website, viewed 20 March 2025, <https://traditionalwatercraftaegeansea.wordpress.com/>

Sabaté, J. and S. Soret 2014. Sustainability of plant-based diets: back to the future. *The American Journal of Clinical Nutrition* 100.1: 476–482.

Smith, K., B. McDonagh and E. Brookes 2024. Place-Based Arts Engagement and Learning Histories: An Effective Tool for Climate Action. *Environmental Communication*: 1–21. <https://doi.org/10.1080/17524032.2024.2382473>

Smylie, M. 2009. *Fishing the European coast*. Cheltenham: The History Press.

Society of Friends of the Museum of Aegean Boatbuilding and Maritime Crafts, Fishing boats destruction, viewed 15 September 2024, <https://woodenboats.gr/en/fishing-boats-destruction/>.

Soilemezidou, M.I. and I. D. Orfanidi 2022. Tourism in the Aegean Sea: Sustainable development and ecology, in D. Barceló and A.G. Kostianoy (eds) *The Handbook of Environmental Chemistry*: 1–28. Cham: Springer.

Throckmorton, P. 1964. Roman shipwrecks and modern Aegean ships. *The Mariner's Mirror* 50.3: 205–216. <https://doi.org/10.1080/00253359.1964.10657778>

Traditional Boat Association of Greece, viewed 15 September 2024, <https://traditionalboats.gr/>.

Trakadas, A. 2021. Maritime Intangible Cultural Heritage: A Role Within the Decade of Ocean Science for Sustainable Development 2021–30. *Intangible Cultural Heritage Courier* 47: 4–7.

Trakadas, A. 2022. The Cultural Heritage Framework Programme: Ensuring a Place for Cultural Heritage's Contribution to the UN Decade of Ocean Science. *Marine Technology Society Journal* 56:110–111. <https://doi.org/10.4031/MTSJ.56.3.29>

Tsartas, P. 2003. Tourism development in Greek insular and coastal areas: Sociocultural changes and crucial policy issues. *Journal of Sustainable Tourism* 11.2-3: 116–132. <https://doi.org/10.1080/09669580308667199>

Tuddenham, D.B. 2010. Maritime cultural landscapes, maritimity and quasi objects. *Journal of Maritime Archaeology* 5: 5–16. <https://doi.org/10.1007/s11457-010-9055-0>

UN Sustainable Development Goals, United Nations, Department of Economic and Social Affairs, Sustainable Development, viewed 15 September 2024, <https://sdgs.un.org/goals> .

Van de Noort, R. 2013. *Climate Change Archaeology: Building Resilience from Research in the World's Coastal Wetlands.* Oxford: Oxford University Press. <https://doi.org/10.1093/acprof:osobl/9780199699551.001.0001>

Velentza, K. 2023. Maritime archaeological research, sustainability, and climate resilience. European Journal of Archaeology 26.3: 359–377. <https://doi.org/10.1017/eaa.2022.48>

Velentza, K. 2024. Traditional shipbuilding on the island of Samos, Greece: Recording the tangible and intangible Data. *International Journal of Nautical Archaeology* 54.1: 129–154. <https://doi.org/10.1080/10572414.2024.2342281>

Virtanen, P. K., L. Siragusa and H. Guttorm 2020. Introduction: toward more inclusive definitions of sustainability. *Current Opinion in Environmental Sustainability* 43: 77–82. <https://doi.org/10.1016/j.cosust.2020.04.003>

Welch, D. 2017. Behaviour change and theories of practice: Contributions, limitations and developments. *Social Business* 7.3-4: 241–261.

Welch, D. and D. Southerton 2019. After Paris: transitions for sustainable consumption. *Sustainability: Science, Practice and Policy* 15.1: 31–44. <https://doi.org/10.1080/15487733.2018.1560861>

Westerdahl, C. 1992. The maritime cultural landscape. *International Journal of Nautical Archaeology* 21.1: 5–14. <https://doi.org/10.1111/j.1095-9270.1992.tb00336.x>

Westerdahl, C. 2011 The Maritime Cultural Landscape, in B. Ford, D.L. Hamilton and A. Catsambis (eds) *The Oxford Handbook of Maritime Archaeology*: 735–762. Oxford: Oxford University Press.

Zacharatos, G.A., M. Markaki, S. Panousi, G. Soklis and A. Christidou 2014. *Η Εποχικότητα του Τουρισμού στην Ελλάδα.* Athens: Research Institute for Tourism.

8. Crafts of the Sea: Documentation of Insular Material Knowledge in the Aegean Sea

Jacob Moe

The Cyclades, an island group in the Aegean Sea, Greece, are home to a rich maritime heritage encompassing wooden boatbuilding, coastal fishing, and seafaring. These practices have existed in Greece throughout ancient and contemporary periods, but currently face significant pressures from overtourism, marine environmental degradation, and climate change. This paper presents the theoretical framework, project methodology and results of *Cycladic Maritime Trades*, a research initiative implemented in 2023-2024 by the NGO Archipelago Network. The project aim was to document the surviving elements of the interlinked endangered material knowledge systems on five Cycladic islands: Amorgos, Koufonisia, Paros, Santorini, and Syros. Drawing on interdisciplinary approaches rooted in archival science, social history, visual anthropology, and maritime archaeology, the project proposes a grassroots community-led approach to documenting and disseminating insular memory and preserving heritage currently under threat. In presenting these methodologies and interpretive frameworks, this paper proposes maritime heritage documentation as a catalyst for pursuing future sustainable development in island communities facing accelerating social and environmental change in the first quarter of the 21st century.

Introduction

Throughout history, seafaring, fishing and boatbuilding have played significant roles in defining the cultural landscape of the Cyclades and have been fundamental to Greece's contemporary maritime identity. Seafaring in the region has facilitated the exchange of goods, peoples and ideas since the Bronze Age and it remains a profession and pastime that is fundamental in reinforcing island communities' relationship with the sea. For millennia, wooden boats have served as connective links between the islands and the surrounding region, and in the modern period have developed into a form of Aegean vernacular architecture internationally recognised for being unique (e.g., Farr 2006; Papageorgiou 2017). Similarly, coastal fishing has long been a key economic activity and driver of social structure in the region, and it encompasses an array of technical and symbolic knowledge concerning the marine environment.

In recent decades, shifting economic, social and technological dynamics have increasingly pushed these knowledge systems to the brink of extinction. Forces such as overtourism and increasingly variable climatic conditions exert additional pressures on these coastal communities. Currently, Greece is experiencing a rapid reduction of wooden boatbuilding and use of traditional wooden boats. As Velentza also notes in Chapter 7, one of the biggest threats faced is the destruction of wooden fishing boats with subsidies provided by the Greek government, in an attempt to implement an EU regulation aimed at reducing the fishing fleet of each EU state and its environmental impact (Papadimitriou, 2019; Velentza 2024). At the same time, initiatives for the safeguarding and dissemination of technical knowledge surrounding these local knowledge systems, for example the Museum of Aegean Boatbuilding and Maritime Crafts and the Wooden Boatbuilding School in Samos, currently in development, offer new paths for training and supporting new generations of traditional boatbuilders.

Within this context, audiovisual documentation of surviving material knowledge relating to traditional seafaring, boatbuilding and fishing in the Aegean is essential for safeguarding, awareness raising, advocacy, and eventual policy changes. Yet the historical record for such material in the Aegean is fragmentary, while scholarly research is underrepresented in the fields of history, sociology and anthropology, especially considering the outsize role played by maritime activity in the historical and social development of modern Greece.

Theoretical Framework

The research initiative *Cycladic Maritime Trades* sought to bridge the gap between academic understanding and practical strategies for the preservation of maritime culture in the Aegean islands of Greece. The project built on previous scholarship of maritime historians and ethnographers on insularity and community-based heritage documentation in the Aegean, as well as the growing body of scientific literature connecting climate change, sustainable development and climate resilience to cultural heritage. In more detail, maritime historians such as Broeze (1987) and Harlaftis (2004) have emphasised the importance of understanding the long-term development of maritime communities and their connections to broader economic and social processes of Greece and Europe. In a detailed study of the ancient Aegean, Constantakopoulou (2007) highlighted how the perception of insularity itself shaped historical realities, influencing trade routes and power dynamics over the long-term history of the region. In the context of modern Greece, Delis (2016) has explored the evolution of maritime trade networks and their impact on island communities such as Syros in the 19th and early 20th centuries, while Alevizakis (2024) provided an overview of marine fisheries from the foundation of the modern Greek state in 1830 to 1910, engaging with a range of previously unpublished historical source material.

Maritime ethnographers such as Olympitou (2025) and Tsimouris (2021) have highlighted the importance of local knowledge systems and practices in shaping maritime communities, ranging from the sponge divers of Kalymnos in the Dodecanese to the sailors on transatlantic cargo vessels. Clay and Olson (2008) have also emphasised the need to understand fishing communities as complex social-ecological systems, with multiple interconnections between human and natural elements. These perspectives are crucial for understanding how climate change impacts not just the environment but also the social fabric of these communities. Research on community engagement in heritage, such as that by Anagnostopoulos, Kyriakidis, and Stefanou (2022), provides insights into how local communities can be actively involved in documenting and preserving their cultural heritage recommending it as best practice. Finally, from a more technical perspective, the contributions of Damianidis (1986; 1991; 1998; 2023) provide a basis for understanding the practical and symbolic knowledge of vernacular boatbuilding construction methods in the Aegean.

The *Cycladic Maritime Trades* project focused on community engagement and participatory documentation. This aligned with recent trends in heritage studies and community-based natural resource management (e.g., Brosius *et al.* 1998). By involving local stakeholders and members of the public in the process of archival digitisation and ethnographic documentation, the project has contributed to raising awareness, building local capacity for heritage preservation and strengthening community ties to maritime traditions. However, it is important to acknowledge the limitations of this approach, including the risk of uncritically accepting heritage narratives without considering how they may be mobilised for various political or economic interests in the present and future.

Methodology

Archipelago Network's mission is the preservation and dissemination of endangered material knowledge in the Aegean Sea region in Greece. Our work is primarily concerned with the creation of photographic and audiovisual archives, media not systematically digitised, analysed and preserved in Greece.[1] The focus that the Archipelago Network has on this media, determines the chronological scope of the organization's research projects, spanning from the advent of photography in the Aegean during the late 1800s up until the so-called 'born digital' imaging technologies of the present day.

In Greece, existing photographic and audiovisual material concerning maritime communities of the Aegean is scattered between various bodies such as public archives (e.g., ERT - Hellenic Broadcasting Corporation archive, and GAK - General State Archives of Greece), local organisations (e.g. local maritime museums, folklore museums, local community associations) and private collections belonging to collectors, craftspeople and other individuals dispersed throughout the Aegean and Greece. Due to a lack of digitisation infrastructure and insufficient expertise in archival documentation, copyright and digital access methods, much of this material is not available for scholarly research or general reproduction. Of the material that is currently available, little is under the open-access licenses that are now becoming standard practice across Europe's GLAM (Galleries, Libraries, Archives, Museums) sector.

To address this fragmentation of the Aegean maritime archival record, *Cycladic Maritime Trades* identified, digitised and published a variety of audiovisual archives from the region. The project primarily focused on historic and contemporary communities of boatbuilders, fishers and captains across the Cyclades, and, therefore, the team conducted on-site fieldwork including video interviews with participating knowledge holders, audio field recordings of boatbuilding and fishing environments, photographic documentation of these contexts, and photogrammetry of selected boatyards and tools. The overarching aim has been to develop a comprehensive framework for community-based cultural documentation rooted in the present-day needs of maritime communities.

The identification of target communities in the Cyclades set off through a combination of desk-based research, informational interviews with contacts throughout the Cyclades, and site visits to meet with local stakeholders. Of the approximately 20 major inhabited islands under consideration, five islands (Amorgos, Koufonisia, Paros, Santorini, and Syros) were selected (Figure 8.1) based on their historical and contemporary importance to boatbuilding and coastal fishing in the region, but also due to the availability and accessibility of relevant material. Each of these islands provided an opportunity to develop unique case studies: Amorgos was selected primarily due to the presence of an active professional fishing association championing sustainable methods of fishing; Koufonisia for its historically significant boatyard, a lifeline for fishers in the 'small Cyclades' islands throughout the 20th century, and its fishing fleet, one of the largest per capita in the Aegean; Paros for its two active boatyards and associations promoting maritime heritage and traditions to youth groups; Santorini for the shuttered but well-preserved traditional boatyard of Armeni as well as for its fishing community, pioneering models for sustainable fishing via fishing tourism; and Syros because it is a historic center of wooden boatbuilding in the Aegean and home to five active boatyards.

[1] It is only recently that an open access platform SearchCulture.gr (the Greek National Aggregator of Digital Cultural Content) has become available. It functions as a general depository of digitised collections from various Greek institutions.

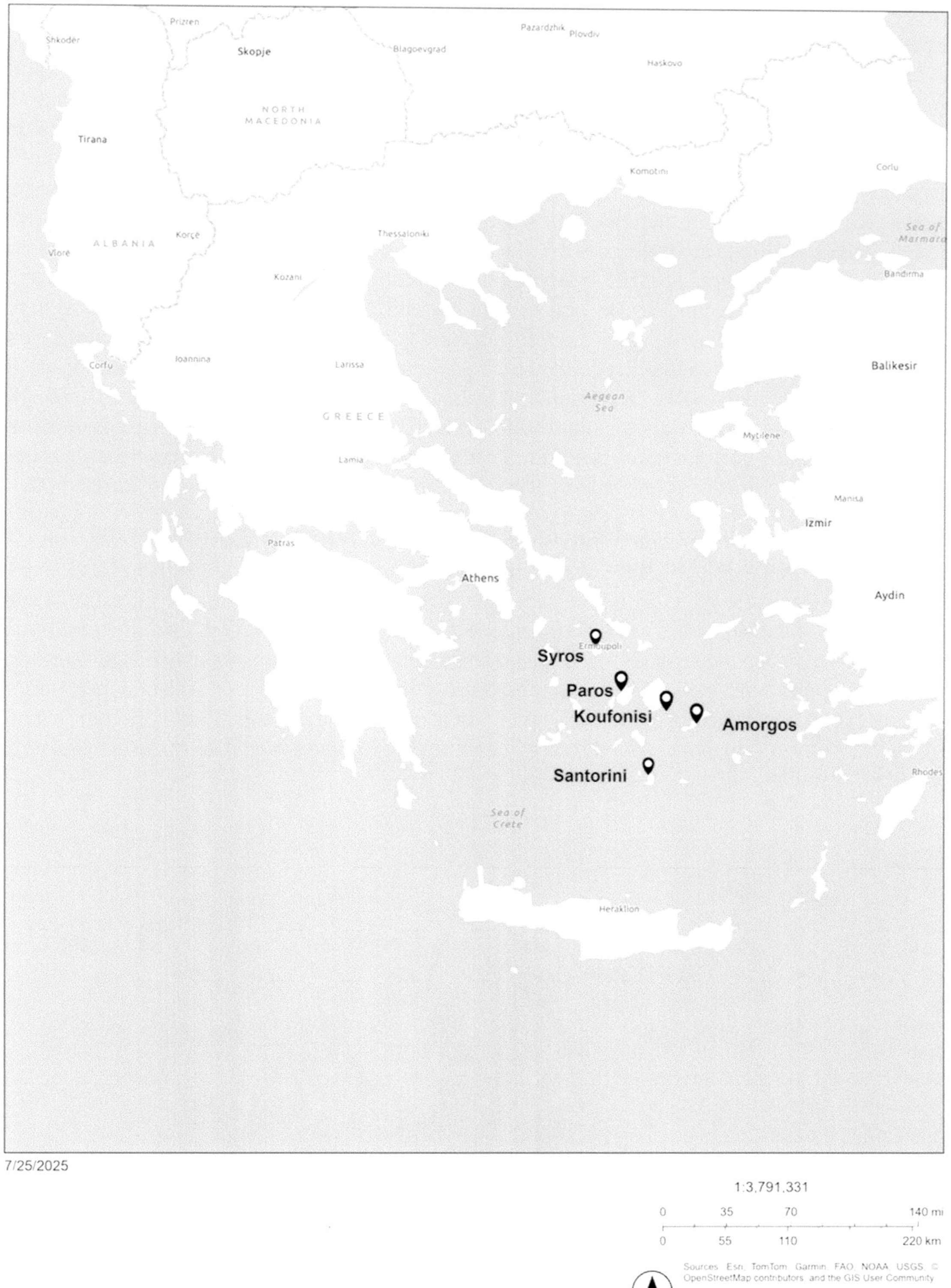

Figure 8.1: Map of island case studies in the Aegean Sea. Copyright: Katerina Velentza.

In parallel to this selection of case studies, methodologies for the project implementation were developed with a steering committee composed of scholars in the fields of maritime archaeology, visual anthropology, and naval architecture. With input from this committee, extended questionnaires were drafted for each of the target communities: boatbuilders, fishers and captains. For example, the questionnaire developed for boatbuilders included four sections that cover a combination of qualitative and quantitative data: boatbuilding activity, biographical/historical data, broader context, and sustainability/policy suggestions. In more detail, the questionnaire's first section covered typologies, morphologies, tools, practical design issues, boatbuilding carpentry knowledge, and a step-by-step discussion on the boat construction process. Biographical/historical data included questions regarding the boatyard itself, in addition to some basic statistical figures (i.e., total number of boats built, size and types of boats, etc.). The next section focused on symbolic knowledge related to the marine environment, as well as the social and economic dimensions of boatbuilding. Finally, the sustainability/policy section requested input from participants concerning their familiarity with sustainable practices and policy suggestions beneficial to their profession. Drawing on participatory ethnographic methodologies, these questionnaires were treated more as points of departure for conversation than fixed scripts, with the content of the discussions being primarily driven by the interviewees during the semi-structured interviews that we carried out.

Having identified the target communities of the project and finalised preliminary research preparations, the Archipelago Network team commenced archival acquisition and on-site documentation in collaboration with the project's partners on each island: individual collectors, cultural/trade associations, and others (Figure 8.2). Archival digitisation and documentation followed a series of procedures established by our team, including a comprehensive legal framework grounded in an open access philosophy, and metadata collection guided by established international archival protocols. Developed progressively over the course of the organisation's previous archival initiatives, this comprehensive archival strategy anchors digitisation activities on a sustainable and ethical basis (Markellou 2024).

Figure 8.2: Fishermen in Ermoupolis port, Syros, c. 1925. Copyright: G.H. Mate collection via Archipelago Network.

As a first consideration regarding digitisation of archival collections, Archipelago Network ensures that the digital archival material it produces shall be managed in accordance with: (a) any legal undertakings given to the depositors; (b) existing copyright legislation; (c) data protection legislation; (d) freedom of information legislation; (e) local cultural heritage legislation. The rights clearance process for these materials strikes a balance between respect for the law and respect for rights holders, while also seeking to meet aims for universal access set out by the International Council on Archives (n.d.). In this process, the following questions have been considered: Who are the archive collection holders? What is the nature of the cultural material? Is there existing copyright protection on this material? What is the age of the material? What is the purpose it was originally created for? If the material is copyright protected, do the archival holders own the copyright of the material? Are there additional legal restrictions that arise from any other kind of legislation that should be seriously taken into consideration? A final consideration is the effective communication of the project's aims to rightsholders and getting permission to use their material. Subsequently, to seek permission and obtain a license, the organisation clarifies potential future uses of the material, emphasising the specific and detailed ethics policy that is followed. Following rights clearance, the project adheres to metadata collection practices outlined by the Material Culture Ethnography Metadata Schema (MCEMS), which provides a formal definition of metadata elements tailored to 'material culture-focused projects recording cultural practices, material knowledge and living heritage [...] which aims to facilitate the inclusion and preservation of metadata for project creating primarily digital recordings, documentation, datasets and archives and provides the potential to connect these to museum collections' (Petek-Sargeant 2020).

Figure 8.3: Screenshot from the digital map and database of the 'Cycladic Maritime Trades' project. Copyright: Archipelago Network.

Once archival materials are sourced and digitised, metadata is recorded using a customised template for each archive at the item level, which combines elements of the MCEMS model with a series of fields compatible with Europeana (n.d.), Europe's portal for cultural heritage, and EKT (National Documentation Center of Greece), the nation's cultural heritage aggregator (n.d.).[2] A selection of the data collected and digitised as part of the project, have been made available to view and search through the project's digital map (Figure 8.3) on the Archipelago network website, a platform that ensures the public dissemination and open-access availability of the data. These collections, supplemented on a regular basis, are made available under a Creative Commons open access license, with the aim of facilitating their circulation and re-use of this material for research, advocacy or preservation of local traditions and heritage.

Results

This research into historic and contemporary heritage data through available archives has provided a multimedia documentary portrait of maritime communities in the Cyclades, offering insights into complex social-ecological systems that characterise these islands. The interdisciplinary approach we followed, combining archival research, ethnographic fieldwork, and audiovisual documentation, allowed for a nuanced understanding of the challenges and opportunities that these communities face.

Overall, 19 archival collections were digitised, encompassing over 450 individual images, audio and video formats. Historic photographic collections depicting fishing activity between 1920-1990, interviews with boatbuilders from the 1980s, and filmed documentation of boatbuilding activity in the 1990s, were included among others. The contemporary photographic documentation was carried out on the five island case studies by the co-investigator of the project Maurizio Borriello. This resulted in 43 discrete thematic photographic collections which depict technical, social, cultural and environmental aspects of boatbuilding, small scale fishing and seafaring, including workshops, boatyards, fishing expeditions, and specialised equipment. Photogrammetry was deployed to create orthomosaic photographs of boatyards and boat maintenance sites on each island, and 3D models of selected woodworking tools in Syros (Tarsanas shipyard) and Santorini (Armeni shipyard).

Koufonisia

Documentation on Koufonisia focused on the former and current traditional boatyards on the island. To record those, we carried out drone photography and video (Figures 8.4-8.6), underwater photography and video (Figure 8.7), and partial digitisation of the photographic archive of Antonis Kovaios, leader of the local municipal community. We also conducted video interviews with locals (Figure 8.8) during which various cultural understandings of the marine environment were underscored, while also highlighting the vulnerability of local knowledge systems to disruption. This is because in such traditional contexts, the perpetuation of knowledge across generations is often hindered by the inherently guarded or private nature of professional expertise. In an interview Kostas Prasinos, a fisherman from Koufonisia, comments on the value of such secrecy: 'The fisherman has to be a liar. If you tell the truth, you're ruined. If you go around telling everyone where you got your big catch... you won't find anything there the next day. They'll have taken everything. If you hear a fisherman telling the truth, then they're no fisherman.' Although such guardedness might have safeguarded resources and livelihoods for local fishermen, in an age when local maritime professions are rapidly disappearing, we found it useful to discuss ways of developing new contexts where exchange and dissemination of selected trade-based information could take place.

[2] This step ensures maximum compatibility of the project's archival materials with wider networks of cultural heritage and inclusion in international repositories and research results.

Figure 8.4: Panoramic view of Koufonisia former boatyard and slipway, April 2023. Copyright: Archipelago Network.

Figure 8.5: Aerial view of current Koufonisia boatyard. Copyright: Archipelago Network.

Figure 8.6: 3D model of current Koufonisia boatyard. Copyright: Archipelago Network.

Figure 8.7: Documentation of Koufonisia boatyard activity, April 2023. Copyright: Archipelago Network.

Figure 8.8 Interview with fisherman of Koufonisia, April 2023. Copyright: Archipelago Network.

Amorgos

Documentation on Amorgos focused on fishing communities of the Katapola and Aegiali ports, as well as the boat maintenance site in Kalotaritisa bay. We used mostly conventional photography as well as drone photography and video to document the sites during our fieldwork. Photographs and documents were also digitised from six personal collections belonging to members of the Amorgos Fishing Association, while extensive visual documentation of fishing equipment and tools (Figures 8.9-8.10) was also conducted. In an interview, the third-generation captain and fisherman Nikolaos Vekris demonstrated the importance of local ecological knowledge in understanding how the sea functions as an aquatic environment: 'The sea is not a meadow. There are islets that we don't see, there are plains we don't see. There are plateaus in the sea. Hundreds of them in the Aegean Sea and the Mediterranean Sea. Sea currents are like the Danube, moving at tremendous speed and, depending on the circumstances, in different directions and at different speeds. Large volumes, millions of cubic meters. It takes a lot of calculations. You can't do it all with a computer. It's all a matter of experience.' This local ecological knowledge in the age of the climate emergency is invaluable because it can allow easier climate adaptation and community resilience in the face of the rapidly changing coastal and maritime climates. By recording this type of embodied and situated observations, gained over decades of navigating the seas, the aim has been to place such knowledge in conversation with established scientific research on climate change and oceanography.

Figure 8.9: Documentation of fishing tools in Katapola, Amorgos. Copyright: Archipelago Network.

Figure 8.10: Documentation of fishing tools in Katapola, Amorgos. Copyright: Archipelago Network.

Paros

Documentation in Paros focused on the fishing communities of Naousa (Figure 8.11), the boatyard of Agios Ioannis, and a wooden cargo boat. During this fieldwork we documented with conventional photography particular types and methods of fishing local to Paros (Figure 8.12). Moreover, drone photography was used to produce orthomosaics and 3D models of the Agios Ioannis boatyard. Interviews with boatbuilder Stathis Aliprantis reveal the kinship-based networks and processes through which vernacular boatbuilding methods are passed along. Aliprantis said: 'for marking the boats, we had these typical red pencils. My ears are a little pointy, so the pencil didn't sit on them. So, I used the pencil I had for school. One pencil is wider and the other more narrow. I started marking the ship with my pencil and my grandfather was marking it with his pencil from the other side... but when our pencils met, the lines weren't aligned. So, the boat would have ended up crooked, obviously. But my grandfather didn't say anything til I had marked all of the boat's main frames... making mistakes, that's how you learn.' In a different interview, Manolis Zisimopoulos, captain of the region's last remaining wooden cargo boat, draws a parallel between cargo trade and the broader socioeconomic transformations occurring in the region: 'As far as I can remember, we used to trade wine. It was exported from Santorini and transported to Piraeus. Or Paros or Volos, those were usually the routes. We also carried barley. We usually brought it to Fix [a Greek beer company], in Athens, to be made into beer. This all happened until 1977. Then gradually, tourism grew. These ships used to bring food to the islands. Food, flour, anything like that. And then they moved on from food... and started transporting materials for construction, cement, bricks etc.' Though his cargo boat used to make dozens of trips in the region, he now only makes one or two a year.

Figure 8.11: Fishermen in Naousa harbour, Paros, c.1965. Copyright: Historical-Folklore Museum of Naousa via Archipelago Network.

Figure 8.12: Net fishing off the coast of Naousa, Paros. Copyright: Archipelago Network.

Figure 8.13: Coastal boat storage units by Armeni, Santorini. Copyright: Archipelago Network.

Santorini

Documentation in Santorini focused on the historic Armeni boatyard, where we carried out photographic recording (Figure 8.13), and on a family of fishers from Vlychada bay working with fishing tourism and sustainable fishing methods. In one of the interviews, Anthi Arvaniti, a fisherwoman from this family, explained how intergenerational knowledge taught her to respect the sea, informing her own work in sustainable fishing methods and fishing tourism. Arvaniti said: 'when I was a child, I remember my grandfather teaching us– below our house, there's this beach called Kalami, filled with stones and rocks. There he taught us that the little shells had to be thrown back into the sea, when they weren't big enough to be harvested. Or that you had to throw the little fish back into the sea, since it wasn't the right time to catch them. This mentality was passed on to his children. My uncles and I took it for granted, as children, that this is what must be done.'

Figure 8.14: Single mold construction method for main frames of boat, Syros. Copyright: Archipelago Network.

Syros

In Syros, photographic documentation and interviews were conducted at all five active boatyards (Figure 8.14), while photogrammetry of the Tarsanas boatyard and selected tools from its workshop resulted in selected orthomosaic photos and 3D models. Among other topics, interviews focused on the interrelation between individual craft workshops and broader economic shifts threatening wooden boatbuilding. Thanasis Vlamis, a boatbuilder from Syros, commented on shortages of specific materials, and how they reflect a broader crisis: 'There are so few of us left, it's hard to imagine a future. I'll give you an example from a boat I'm working on now. I'm caulking the boat now, and the simplest thing–the cotton yarn we use as material to caulk the boat's seams–isn't produced in Greece anymore. In other words, to get the most basic things, one needs to go to Turkey [...] Everything vanishes, bit by bit.'

The 'Crafts of the Sea' traveling exhibition

The above examples of documentation give a glimpse of the local ecological knowledge and community resilience that has been associated with the traditional maritime trades and crafts of the Aegean Sea region. These elements still exist with some members of these communities, as empirical understandings and lived experiences, but are at serious risk in the face of changing economic and environmental conditions. This also underscores the vulnerability of these knowledge systems to disruption, particularly as younger generations may be drawn away from traditional maritime practices due to economic pressures or changing aspirations, which could then affect the adaptability and longer-term survival of these communities as a whole. This in turn emphasises the need for active community engagement in preserving this heritage and the local knowledge systems that it carries.

Figure 8.15: Installation view of 'Crafts of the Sea' exhibition at the Goethe-Institut Athen. Copyright: Stathis Mamalakis.

To address that need for active dissemination, we created the '*Crafts of the Sea*' traveling exhibition (Figure 8.15). This was a key component of the project, designed to move beyond a mere presentation of research findings, but engage actively the public of the Aegean with local knowledge related to maritime crafts. The first instance of the exhibition, held in Amorgos in collaboration with the Professional Fishers' Association, involved collaborative curation of the Association's photographic archive, digitised by Archipelago Network. The following iteration, held in Athens at the *Goethe Institut Athen*, involved material from all five islands, as well as a screening series bringing the project's film portraits into conversation with ethnographic and documentary films from maritime communities in other regions around the world (sponge divers in Korea and the Moken people of the Andaman Sea, among others). The final chapter of the exhibition, held in Paros in collaboration with the Paros municipality and the Fishing Association of Naousa, included material from all five islands, with a focus on Paros, and a presentation of film portraits from Paros during the island's Annual Regatta for traditional boats.

Drawing on related scholarship, such as the work of Buck *et al.* (2020), the exhibition sought to create an immersive experience that fostered dialogue and engagement with the documented maritime traditions. The design incorporated oral history testimonies, photographic displays, and film screenings within a modular, custom-built wooden structure that traveled to all locations. The exhibition design considered sustainability by using recyclable materials, minimising shipping costs by engaging local suppliers on each island for materials when possible. Following Day (1999) and Cento Bull and Raynolds (2021) we recognised the importance of ensuring communities retain power throughout the exhibition. The exhibit was explicitly designed as a collaborative space, with all exhibition narrative texts and captions checked by members of the communities involved. The meaning-making process involved carefully selecting and framing the research findings to create exhibits that were both informative and emotionally resonant to local community partners, while highlighting the interconnectedness of maritime traditions, environmental change, and community identity.

Conclusion

Through multi-modal documentation of cultural heritage related to maritime crafts of the Cycladic islands in the Aegean Sea, this project has aimed at raising awareness for the living traditions that are currently at risk, while also supplementing the fragmented archival record of audiovisual material related to maritime heritage. The project's interactive map, along with the *'Crafts of the Sea'* exhibition have been of core importance in the dissemination of the material documented, and also the initiation of a multi-step dialogue that relates to the significance of maritime heritage and traditions in building community resilience and pursuing environmental sustainability in the face of the climate emergency and the associated socio-economic crises of the 21st century.

Future research and policy initiatives could focus on how this documented heritage can be leveraged to support practical solutions for sustainable development in these Aegean communities. This might involve exploring innovative ways to integrate traditional maritime knowledge with contemporary economic opportunities, such as sustainable tourism or artisanal fishing practices that align with ecosystem conservation goals. Moreover, future work could more explicitly address the challenges posed by climate change to these maritime communities, such as extreme storminess causing flooding or sea level rise. While this project touched on maritime environmental knowledge through the lens of the local knowledge of traditional boatbuilding, fishing and seafaring communities, future expansion of this project's results could include a targeted investigation of how these knowledge systems and practices can contribute practically to climate change adaptation strategies.

Finally, this project underscores the need for continued interdisciplinary research that bridges the gaps between cultural heritage preservation, environmental management, and community resilience. By understanding maritime communities as complex social-ecological systems, we can work towards more holistic and sustainable approaches to preserving and revitalising these valuable cultural traditions.

References

Alevizakis, N. 2024. The marine fishery from the establishment of the Greek state until 1910. Unpublished PhD dissertation, Ionian University. <http://hdl.handle.net/10442/hedi/55990>

Anagnostopoulos, P., E. Kyriakidis, E. and E. Stefanou 2022. *Making Heritage Together: Archaeological Ethnography and Community Engagement with a Rural Community*. Abingdon: Routledge.

Broeze, F. 1987. *The maritime history of the world*. Cambridge University Press.

Brosius, J. P., A.L. Tsing and C. Zerner 1998. Representing communities: Histories and politics of community-based natural resource management. *Society & Natural Resources* 11.2: 157-168.

Buck, L., J.H. Gilmore and L. Taurins 2020. *Museum exhibition design*. Abingdon: Routledge.

Cento Bull, A. and S. Raynolds 2021. *Oral history in museums*. London: Bloomsbury Publishing.

Clay, P. M. and J. Olson 2008. Defining fishing communities. *American Fisheries Society Symposium 49*: 7-24.

Constantakopoulou, C. 2007. *The dance of the islands: Insularity, networks, the Athenian Empire, and the Aegean World*. Oxford: Oxford University Press.

Damianidis, K. 1991. Vernacular Boats and Boatbuilding in Greece, Volumes 1 and 2. Unpublished PhD dissertation, University of St Andrews, Scotland. <http://hdl.handle.net/10023/7116>

Damianidis, K. 1998. *Η Ελληνική Παραδοσιακή Ναυπηγική*. Athens: Cultural Technological Foundation of the Industrial Development Bank of Greece.

Damianidis, K. 2023. Μουσειό Ναυπηγικών και Ναυτικών Τεχνών του Αιγαίου: Η προετοιμασία ενός Μουσείου – Εργαστηρίου για την ξυλοναυπηγική στα χρόνια της οικονομικής κρίσης, in P. Karampampas (ed.) *Άυλη Πολιτιστική Κληρονομιά σε Καιρούς Οικονομικής Κρίσης: Ένταξη στην Αγορά και Ανθεκτικότητα*: 121–126. Athens: Greek Ministry of Culture.

Damianidis, K. and A. Zivas 1986. *Το Τρεχαντήρι στην Ελληνική Ναυπηγική Τέχνη*. Athens: Hellenic Organization of Small and Medium Sized Industries and Handicrafts.

Damianidis, K., L. Mendoni, A. Charoni, V. Polychroni, E. Mpechraki and R. Oikonomou 1998. *Σάμος – Παραδοσιακή Ναυπηγική στο Αιγαίο: Ταρσανάδες και Σκαριά*. Athens: Greek Ministry of the Aegean.

Day, J. 1999. *Exhibiting culture: The poetics and politics of museum display*. Washington D.C.: Smithsonian Institution Press.

Delis, A. 2016. *Mediterranean Wooden Shipbuilding. Economy, Technology and Institutions in Syros in the Nineteenth Century*. Leiden; Boston: Brill. <https://doi.org/10.1163/9789004306158_006>

Europeana, viewed 10 October 2025, <https://www.europeana.eu/en>

Farr, H. 2006. Seafaring as social action. *Journal of Maritime Archrchaeology* 1: 85–99. <https://doi.org/10.1007/s11457-005-9002-7>

García-Quijano, C. (2007). Social-ecological mismatches: The case of the spiny lobster fisheries in Bahía Almirante, Panama. *Ecology and Society*, *12*(2).

Greek National Aggregator of Digital Cultural Content, viewed 20 October 2025, <https://www.searchculture.gr/aggregator/>

Harlaftis, G. 2004. *A global history of shipping*. London: Reaktion Books.

International Council on Archives, viewed 23 October 2025, <https://www.ica.org/>.

Markellou, M., 2024, Archipelago Network's IP methodology, email communication.

Olympitou, E. 2025. *Aegean Sponge Fishing and the Island of Kalymnos (19th-20th Centuries)*. Leiden; Boston: Brill.

Papadimitriou, G. 2019. Η καταστροφή των καϊκιών – Το κόστος του Πολιτισμού. Unpublished Master's dissertation, Cross-institutional and cross-departmental programme of postgraduate studies, National and Kapodistrian University of Athens, University of Thessaly and University of Western Attica, Greece.

Papageorgiou, D. 2008. The marine environment and its influence on seafaring and maritime routes in the prehistoric Aegean. *European Journal of Archaeology* 11.2–3: 199–222. <https://doi.org/10.1177/1461957109106374>

Petek-Sargeant, E. 2020. Best practices for digital cultural heritage projects. *International Journal of Digital Heritage* 1.1: 1-14.

Tsimouris, G. 2021. *Εμείς οι ναυτικοί, μπαρκαρισμένοι και ξέμπαρκοι*. Athens: Da Vinci.

Velentza, K. 2024. Traditional shipbuilding on the island of Samos, Greece: Recording the tangible and intangible Data. *International Journal of Nautical Archaeology* 54.1: 129–154. <https://doi.org/10.1080/10572414.2024.2342281>